SpringerBriefs in Materials

Series Editors

Sujata K. Bhatia, University of Delaware, Newark, DE, USA

Alain Diebold, Schenectady, NY, USA

Juejun Hu, Department of Materials Science and Engineering, Massachusetts Institute of Technology, Cambridge, MA, USA

Kannan M. Krishnan, University of Washington, Seattle, WA, USA

Dario Narducci, Department of Materials Science, University of Milano Bicocca, Milano, Italy

Suprakas Sinha Ray ⓘ, Centre for Nanostructures Materials, Council for Scientific and Industrial Research, Brummeria, Pretoria, South Africa

Gerhard Wilde, Altenberge, Nordrhein-Westfalen, Germany

The SpringerBriefs Series in Materials presents highly relevant, concise monographs on a wide range of topics covering fundamental advances and new applications in the field. Areas of interest include topical information on innovative, structural and functional materials and composites as well as fundamental principles, physical properties, materials theory and design. SpringerBriefs present succinct summaries of cutting-edge research and practical applications across a wide spectrum of fields. Featuring compact volumes of 50 to 125 pages, the series covers a range of content from professional to academic. Typical topics might include

- A timely report of state-of-the art analytical techniques
- A bridge between new research results, as published in journal articles, and a contextual literature review
- A snapshot of a hot or emerging topic
- An in-depth case study or clinical example
- A presentation of core concepts that students must understand in order to make independent contributions

Briefs are characterized by fast, global electronic dissemination, standard publishing contracts, standardized manuscript preparation and formatting guidelines, and expedited production schedules.

More information about this series at https://link.springer.com/bookseries/10111

Samson Mil'shtein · Dhawal Asthana

Harvesting Solar Energy

Efficient Methods and Materials Using
Cascaded Solar Cells

 Springer

Samson Mil'shtein
Department of Electrical and Computer
Engineering
University of Massachusetts Lowell
Lowell, MA, USA

Dhawal Asthana
Department of Electrical and Computer
Engineering
University of Massachusetts Lowell
Lowell, MA, USA

ISSN 2192-1091   ISSN 2192-1105   (electronic)
SpringerBriefs in Materials
ISBN 978-3-030-93379-1   ISBN 978-3-030-93380-7   (eBook)
https://doi.org/10.1007/978-3-030-93380-7

This Springer imprint is published by the registered company Springer Nature Switzerland AG
The registered company address is: Gewerbestrasse 11, 6330 Cham, Switzerland

Preface

This book deals with existing technologies of solar energy conversion as well as novel methods, being under consideration in academic and commercial R&D sites. The experimental results presented in the work are well crafted by both analytical and first-principle numerical simulations. The book highlights the real potential for economically justified use of solar energy in every household and/or commercial solar farms. The ever-improving methods of thin-film epitaxial growth combined with a better understanding of the sun light absorption and anti-reflection are highlighted. While there was a period when the material quality was considered to be cornerstone of the conversion efficiency followed by substantial efforts to optimize multiple-cell architecture, it became clear that many old ideas such as variable band gap, multi-junction intrinsic region, as well as solar tracking mechanisms offer new possibilities for improved harvesting of energy. While amplifying the importance of material selection, efficient design of the photo-voltaic elements, various aspects of the production cost, and their impact on the environment are discussed. In addition, the eligibility of the proposed production technologies in the current PV market are evaluated and confirmed.

In summary: Current research and development of photovoltaic technology in recent years address the design of efficient solar cells, methods of increasing solar hours daily and longevity and cost of PV technology.

Lowell, MA, USA

Samson Mil'shtein
Dhawal Asthana

Contents

Chapter 1
Design of Heterojunction Cascaded Solar Cells

Abstract The widely studied heterostructured solar cells are represented by two major groups, namely tandem and cascaded solar cell configurations. The major element in solar cells of both groups is the intrinsic region of the p-i-n junction, where electron–hole pairs are generated by absorbed light and, most importantly, separated one from the other. The energy gap and absorption coefficient of semiconductor used for the i-region define the harvested part of solar spectra and efficiency of the solar cell. Tandem solar cells are built as stacks of separate p–n junctions, using semiconductors with different energy gaps. On the contrary, cascaded solar cells consist of a set of i-regions made of semiconductors with different energy gaps, but built into a single p–n junction.

1.1 Electricity Generation by Clean Energy Sources

Developers of various alternative energy systems are facing the challenge of uncontrollable increase of the temperature of our planet. Excessive production of CO_2 by many countries around the globe and naturally changing angle of inclination of the axis of our planet are major reasons for steady warming up of our planet. More drastically, it is expected that the increase in temperature of our planet might contribute to the rise of sea levels. Decrease of CO_2 production generated by burning coal and fossils could be achieved if the fraction of overall energy production generated by PV and wind turbines would significantly intensify [1]. For example, annual production of electricity worldwide reached 27,000 TW, of which only about 2% is contributed by PV systems. To get a better understanding of the potential of various clean energy generation technologies, one should analyze the cost of investment, called energy invested (EI) and energy return (ER). Analyzing results reported by Prof. M. Fermeglia of Thrieste University (Italy) at MRS seminar on February 10, 2021, we extracted from that presentation [2] some relevant data, which is tabulated in Table 1.1. The ER/EI = 47:1 ratio demonstrates the attractiveness of coal in production of thermal energy, but it is less attractive in production of electricity. Gas as the source for production of thermal energy is supported by ER/EI = 19:1 ratio; however, burning gas for electricity production carries much smaller ER/EI ratio.

S. Mil'shtein and D. Asthana, *Harvesting Solar Energy*, SpringerBriefs
in Materials, https://doi.org/10.1007/978-3-030-93380-7_1

Table 1.1 Depiction of dataset derived from [2] Prof. Fermeglia presentation

Energy	Energy	ER/EI
Coal	Thermal	47:1
	Electricity	17:1
Gas	Thermal	19:1
	Electricity	8:1
PV	Electricity	25:1
Wind	Electricity	18:1

Usage of solar cells for production of electricity is supported by reasonably high ER/EI = 25:1 ratio; however, wind turbines are less productive in electricity generation than solar cells. The sporadic timing of energy production by solar or wind impacts their usefulness and somehow modifies the ER/EI ratios. The fundamental idea of recycling the energy, for example using energy produced by solar cells for synthesis of fuel like gasoline, and other types of energy is described in [1]. This energy recycling idea might help, in our opinion, to mitigate the technical factor known as increase of earth temperature.

Obviously, the ER/EI ratios in Table 1.1 can be improved for PV, if efficiency of solar cells is improved, cost of semiconductor materials is decreased, and, most importantly, cost of production, i.e., cost of epitaxial fabrication technology, would go down. In the current study, we have focused our discussion on technical performance and potential cost of energy produced by two major groups of solar cells, namely tandem and cascaded solar cells.

The cost of solar farms using mass production of simple p-i-n solar cells was reviewed [3] in 2017. The cost of cascaded solar cells (p-i-i-n structures) is not counted yet, as this technology is not in commercial production. The cost of tandem solar cells using III-V semiconductor compounds is identified to be very costly [4]. However, we are not focusing our comparison on cost of the materials, rather on the impact of design complexity and number of epitaxial layers used in typical products of the two groups. At the same time, our comparison report is not suggesting replacement of tandem solar cells by cells with built-in intrinsic segments. Attention paid to the complexity of fabrication technology brings us to the length of time a given production procedure requires and how much is the selected production cost. After all, we are competing with the time factor of the planet warming up.

1.2 Hours of PV Operation Daily. Solar Cells with Corrugated Top

There are at least two ways to increase the number of hours for solar cells to work productively during the day. Installing solar cells on tracking/anti-tracking systems is one of the methods. The other method is to design solar cells with corrugated surfaces. Below we discuss the design of silicon heterostructure solar cells with corrugated surfaces.

There are several major concepts of p–n junction interaction with the solar energy spectrum, which define the efficiency of a solar cell; we will list them in a sequence of importance from high to low. First, successful harvesting of solar energy could be performed by the devices based on materials which combine various energy gaps: wide E_g at the top of the device and decreasing E_g toward the bottom [5, 6]. This sequence is created to prevent overheating of the solar cell by photons with high energy. It is known that semiconductor solar cells lose 1% of efficiency per every 10 °C of heating up as far as solar energy conversion is concerned. The most important part of the cell design is an intrinsic region of a p–n junction, where recombination of photoelectrons and photo-holes is minimized. Transport and collection of electron–hole pairs are mostly controlled by the Fermi energy profile of the solar cell so the modeling of energy diagrams should be free from potential barriers preventing flow of electrons and holes to the terminal contacts. In the current study, we present the design of efficient solar cells, where harvesting of solar energy, high rate of generation of electron–hole pairs and maximum collection efficiency are achieved.

The described solar cell is not expected to be installed on a sun tracking system. The prolonged illumination per day of the solar cell is provided by special corrugation [7] of the surface of the device and by use of two anti-reflective layers. We present the design of the heterostructured solar cell made of two materials, namely a-Si (amorphous) and c-Si (crystalline), where diode configuration p-i-i-n is used. The top intrinsic areas are made of amorphous Si with $E_g = 1.9$ eV followed by an intrinsic crystalline region with $E_g = 1.12$ eV. We also discuss the sequence of fabrication steps which allow manufacturing our novel solar cell.

The proposed structure consists of 1 μm thick 10^{16} cm^{-3} n-doped c-Si at the bottom, on top of which there are intrinsic c-Si and a-Si regions consecutively (2.5 μm thick each), followed by 0.1 μm thick 10^{16} cm^{-3} p-doped a-Si, 0.07 μm Si$_3$N$_4$ and 0.01 μm SiO$_2$ (see Fig. 1.1).

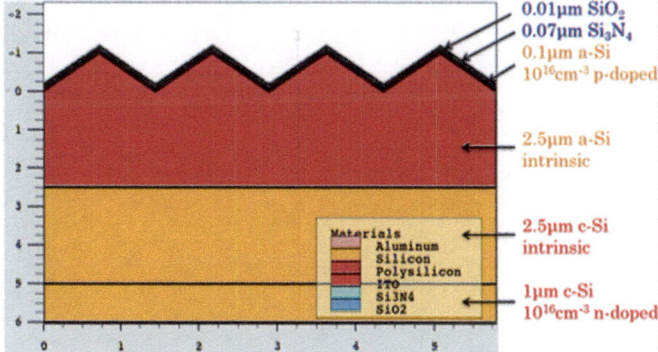

Fig. 1.1 p-i-i-n solar cell structure with corrugated surface [8]

Application of a very thin acceptor-doped region is needed to diminish the photo-generation rate in that region. This allows more photons to pass through to the more efficient intrinsic regions where the electron–hole pairs' recombination rates are low. Intrinsic regions are thicker than doped layers as that allows creating more electron–hole pairs in the region. Separation of electron–hole pairs and collection of photo-carriers are assisted by the internal electric field throughout the depleted region. Ohmic contacts to n-type and p-type regions are provided by 20 Å of heavily doped (10^{20} cm^{-3}) n$^+$layer and 20 Å of 10^{20} cm^{-3} p$^+$ layer.

The importance of a-Si and c-Si cascaded structure is defined by the difference between the energy gaps of these two materials. Having a-Si with the energy gap of 1.97 eV precede c-Si with 1.12 eV gap on the solar beams' path allows the structure to harvest high-energy photons and not allow them to overheat the c-Si region. This positively affects the lifetime of photogenerated electron–hole pairs in the intrinsic crystalline Si region, increasing the overall efficiency of the structure. Corrugated surface of the solar cell allows increasing the light trap properties of the structure when the sun is low above the horizon without using high-cost solar tracking systems. Our proposed solar cell structure has a corrugated p-doped and amorphous crystalline i-region with 1 μm high and 1.452 μm wide pyramids (to achieve optimal 54° angle [7]). These were covered by 0.07 μm Si$_3$N$_4$ and 0.01 μm SiO$_2$ double-layer anti-reflection coating (DLARC). DLARC's thicknesses were optimized [9] to have an overall lower reflection level within 450–1100 nm wavelength range (refer to Fig. 1.2). Light source specifications in the simulation of the structure were set up so the source structure, so the source emits light in a 400–1100 nm wavelength range. Efficiency calculations were adjusted for the same range.

The Silvaco modeling software, which was used for modeling, is briefly described below. The simulation method applied to this structure was ray tracing 2D in Silvaco device simulation software. This method takes into consideration reflection and changes in speed/direction of light in the different layers of the structure; it does not, however, account for interferences of the waves in the DLARC region, and therefore, the thicknesses of the DLARC was set in order to optimize the physical model only.

Fig. 1.2 (x-axis: wavelength in nm, y-axis: percentage reflection) DLARC reflection for 0.01-μm-thick SiO$_2$ and 0.07-μm-thick Si$_3$N$_4$

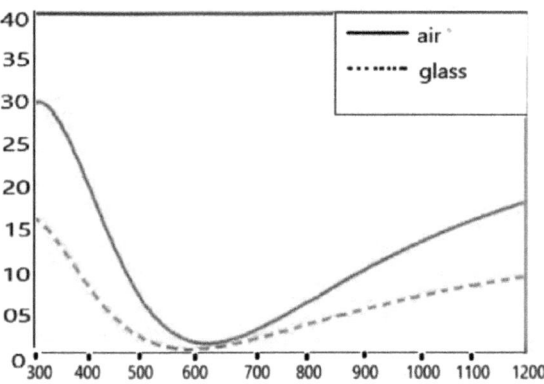

The modeling and simulation of this structure were performed using Silvaco DevEdit and Atlas TCAD [10] (version 5.20.2.R) software, which provide a wide range of tools for process and device simulations. This section briefly covers some important aspects of the device simulation process.

Defining an adequate set of grid nodes of the structure, meshing, is crucial since otherwise it would lead to incorrect simulation of a device and consequently to inadequate results. Finite element analysis is the base of Silvaco model simulations, where high accuracy of grid nodes is required.

When the physical structure of the solar cell is defined, the material parameters are introduced into modeling. We started with the refractive coefficients for DLARC and reflective back contact which are specified by defining the anti-reflection index using interface statements. Then the recombination lifetimes are specified for each of the regions. In our structure, the lifetimes used were 10^{-6} s for p-type a-Si; 10^{-5} s for i-type a-Si; 10^{-3} s for i-type c-Si, and 10^{-3} s for n-type c-Si. The back reflection parameters were defined.

Figure 1.3 represents generation/recombination rates of electron–hole pairs throughout the modeled structure. As expected, the generation of photo-carriers is very high in every layer of the device. Recombination rate in intrinsic c-Si is low; however, in intrinsic a-Si, recombination rates are increased due to the large density of electron dangling bonds, i.e., large density of defects.

Figure 1.4 shows the energy band diagram of the structure with doping information about the layers in our device. The left side of the diagram corresponds to the top, p-type part, of the solar cell and the right side to the n-type bottom part of the cell; dashed vertical lines in the middle represent the intrinsic segments of amorphous and crystalline interface.

Figure 1.5 and Table 1.2 contain solar cell simulation results. The modeling describes a very small cross section of the solar cell, i.e., 2.904 μm^2.

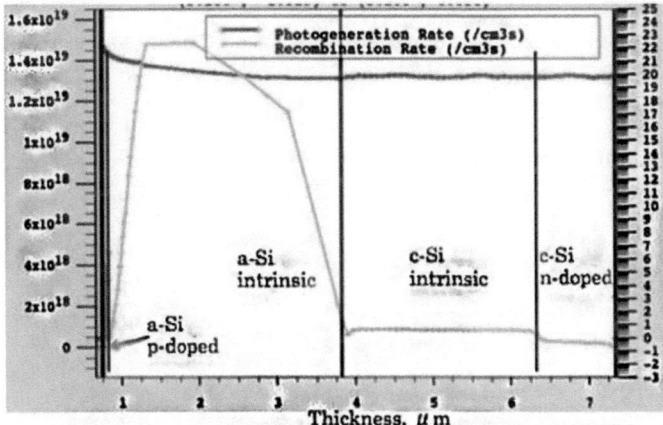

Fig. 1.3 Photogeneration and recombination rates [8]

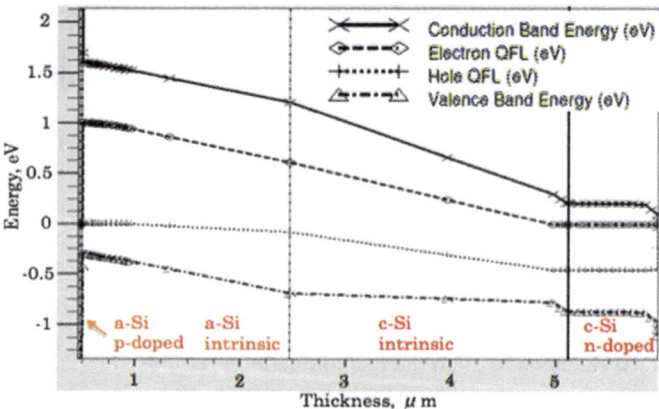

Fig. 1.4 Energy band diagram of the a-Si/c-Si structure [8]

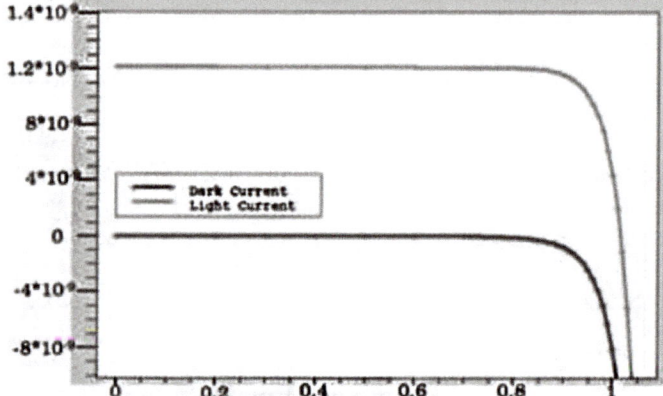

Fig. 1.5 I-V characteristics of the structure [8]. (y-axis: Output Current (A), x-axis: Anode Voltage (V))

Parameter	Value
J_{sc}	21.4798 mA/cm^2
V_{oc}	1.0843 V
P_{max}	$9.52709 * 10^{-10}$ W
V_{max}	0.979995 V
I_{max}	$9.72157 * 10^{-10}$ A
FF	0.869906
Efficiency	29.4855%

Table 1.2 Major performance parameters of a-Si/c-Si corrugated

1.3 Proposed Fabrication Steps for a-Si/c-Si Cascaded Solar Cell

The difficulties to formulate the processing steps of amorphous layers and prescribing precise plasma parameters and/or annealing temperatures come from the fact that atomic structure of amorphous Si does not follow the rule of four covalent bonds per atom. Even hydrogenated Si has at most 2.7 bonds per atom [11, 12]. Random distribution of the density of dangling bonds in amorphous Si significantly complicates the control of doping, especially donor doping. This is described in many studies and repeated in review [11]. On a macroscale, presence of voids in this amorphous material undermines uniformity of interfaces of a-Si/c-Si, which is critical for design of semiconductor junctions. Finally, the degradation phenomena caused by Staebler–Vronski effect [12] require certain modifications of fabrication procedures [13]. Although the mentioned above negative effects are well studied and documented, the attempts of making a-Si/c-Si heterostructure solar cells are motivated by low cost of materials and production steps [14] and by maturity of silicon technology.

We would like to suggest the following fabrication steps of our novel cascaded a-Si/c-Si p-i-i-n solar cell:

(a) Thin layer n-type c-Si could be produced by chemical vapor deposition (CVD).
(b) Thin intrinsic layer of c-Si could be produced by the same technology.
(c) Corrugation profile is created by selective etching of intrinsic c-Si layer. The position of top amorphous layers would replicate the corrugated profile.
(d) Thin intrinsic and doped layers of a-Si would require enhanced plasma deposition (EPD) of a hydrogenated selection.

To summarize our modeling, we can state that the high efficiency of 29.5% a-Si/c-Si solar cell was designed, where cascade i-i layers are built within the same p–n junction. This is the major difference from the design of multi-junction solar cells. Corrugated surface increases the number of sun hours by 30%. Random distribution of the density of dangling bonds in amorphous Si impacts the solar cell parameters. This is the reason why commercial companies have their own manufacturing recipes and that is why we suggest above only a tentative sequence of fabrication steps.

1.4 Low-Cost Heterostructure p-i-i-n CdS/CdSe (Cascaded) Solar Cell

Dominant presence of silicon in solar cells manufacturing is defined by maturity of this technology and low-cost of thin-film solar cells made from this material. Better efficiency of cascaded solar cells compared to tandem devices performance was demonstrated in numerous uses of our recent publications [6–8, 15–18]. Successful development of efficient solar cells is built nowadays on a combination of harvesting of a wider range of solar energy and design of multi-semiconductor layers with a viable energy gap. Thus, most of more sophisticated solar cells are using either tandem or heterostructure designs. The success of different semiconductor materials is well reflected in recent Fraunhofer market study [19]. According to the Fraunhofer report [19], the II-IV semiconductor compounds such as CdS, CdTe and/or CdSe started recently to prevail in competition with silicon (see Fig. 1.6) mostly due to the low cost of thin films of these compound semiconductors. The tendency of increased production of CdS and CdSe continues in the last decade. The quantum dots made of CdS and CdSe already demonstrated 10–17% efficiency [20]. The development of CdTe/CdS solar cells on flexible substrates with efficiency of 11% was reviewed recently [21]. Photovoltaic (PV) solar cells based on cadmium telluride (CdTe) represent the largest segment of commercial thin-film module production worldwide. Recent improvements have matched the efficiency of multi-crystalline silicon while maintaining cost leadership. The USA is the leader in CdTe PV manufacturing, and NREL [22] has been at the forefront of research and development (R&D) in this area. In Europe, Fraunhofer Research Institute demonstrated in joint German–French development of tandem SOI solar cells with concentrator's efficiency of 46% [19]. The efforts to improve manufacturing technology of CdS and CdSe solar cells continue [22].

The deficiency of Si is defined mostly by low absorption coefficient [23], which requires usage of thick layers of the material. The relatively small energy gap of Si leads to significant overheating [24] of the solar panels, which in turn decreases the operational efficiency of these widely used panels. Efficiency of silicon-based solar cells was significantly improved by design of a heterostructure of a-Si/c-Si. Research groups across the world [11, 25] did prove that the usage of wide gap amorphous silicon provides much higher efficiency compared to polysilicon or single crystal silicon solar cells.

Our design of heterostructure thin-film silicon solar cells was different from any other design [13, 26]. Offering the heterostructure design with novel configuration p-i-i-n, we secured light absorption and generation of photo-carriers in two intrinsic regions, where recombination of these carriers is significantly suppressed. The top

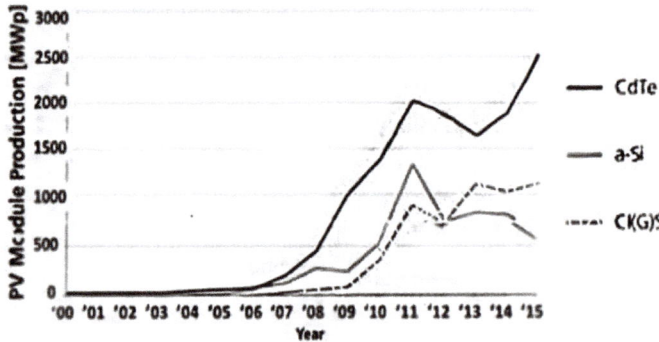

Fig. 1.6 Annual global PV production from 2000 to 2010 (after [19])

p-region was designed to be very thin. Similar design ideas are explored in our novel CdS/CdSe p-i-i-n solar cell.

Comparing results of our recent design [15] of CdS/CdSe 15.6% efficient solar cell with cascaded p-i-n configuration, we tend to believe that 18.5% efficiency of heterostructure CdS/CdSe p-i-i-n solar cell is achieved due to the presence of double intrinsic layers. In current design, we suggested the necessity to insert a buffer layer between the intrinsic regions to match lattice constants of CdS and CdSe. The cost efficiency of novel design is also assessed. The solar cell (see Fig. 1.7) consists of 0.1-μm-thick acceptor-doped $(2 \times 10^{18}\,\text{cm}^{-3})$ CdS layer followed by 2 μm intrinsic CdS. Below is intrinsic CdSe 2 μm thick followed by 0.1 μm thick donor-doped $(2 \times 10^{16}\,\text{cm}^{-3})$ CdSe. Energy diagram is presented by Fig. 1.8. The slope of intrinsic regions demonstrates favorable conditions for collection of both photoelectrons and photo-holes.

Fig. 1.7 Structure of Cds/CdSe p-i-i-n solar cell [17]

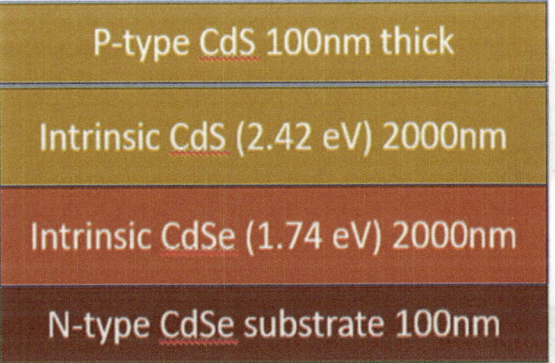

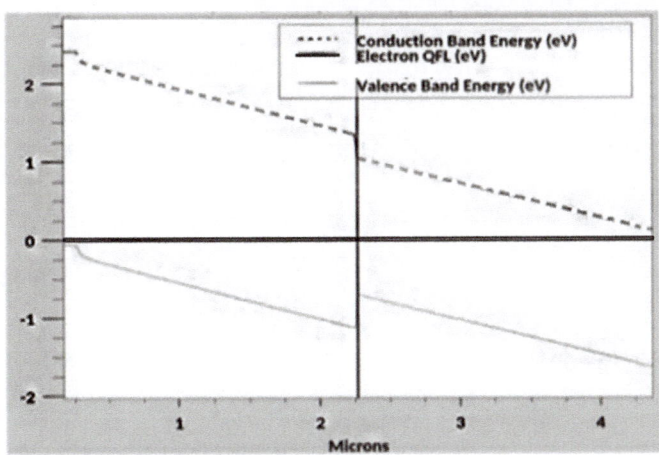

Fig. 1.8 Energy band diagram of CdS/CdSe cascaded p-i-i-n solar cell [17]. (y-axis: energy level in eV)

Design choices of the region thicknesses are described in this section. Optimal thickness configuration was expected to have a relatively smaller p-region, and this is since all electron–hole pairs created in that region will immediately recombine unlike inside of a p-i-i-n junction where carriers will be pulled apart by the built-in field. To maximize the light absorption, it is advantageous to have intrinsic regions relatively thick. If they are too thick, however, the strength of the built-in field decreases, in turn decreasing efficient harvesting of solar energy. The modeling indicated that the optimal thicknesses (see Fig. 1.7) for the regions are 0.1 μm for p-doped CdS and 2 μm for both intrinsic CdS and CdSe. Our calculations with the extremely small lifetime of carriers in CdSe [8] motivated us to use the thickness of 0.1 μm for n-doped CdSe to decrease electron–hole recombination in that region and make the n-region smaller than the diffusion length, while keeping it thick enough for potential CVD manufacturing processes.

Design choices regarding the doping configuration were governed mainly by a need to have built-in potential across the junction as large as possible, while keeping the recombination of majority and minority carriers in p and n-regions as small as possible to achieve maximum efficiency. Our modeling showed that having 2.3×10^{18} cm^{-3} of acceptors in p-doped CdS region with 10^{16} cm^{-3} donors in n-doped CdSe region provides optimal solar cell efficiency. In this simplified model, it was assumed that the regions are doped uniformly.

Figure 1.9 shows the I-V curve of the designed cell. Note that the model cross section of the device was said to be 1 μm^2, so the current values shown on the graph should be converted into current densities to be compared to performance of other solar cells.

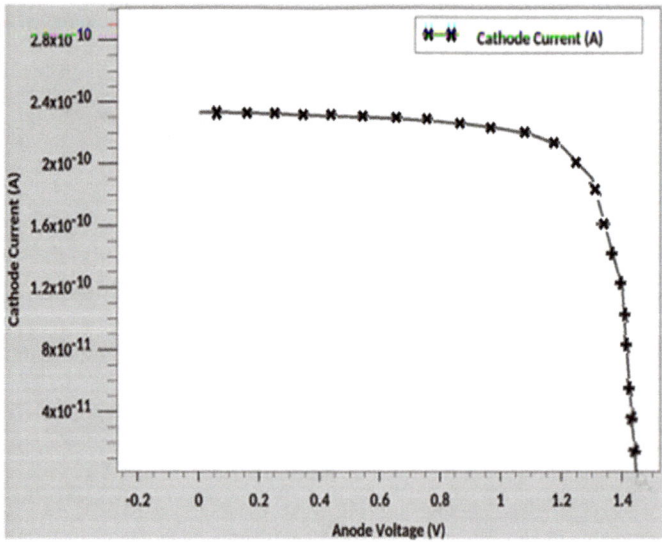

Fig. 1.9 I-V relationship of the CdS/CdSe cascaded p-i-i-n solar cell [17]

Table 1.3 CdS/CdSe p-i-i-n solar cell characteristics [17]

J_{sc}	23.36 mA * cm^{-2}
V_{oc}	1.45 V
P_m	25.37 mW * cm^{-2}
V_m	1.200 V
I_m	21.15 mA * cm^{-2}
FF	74.99%
E$_{ff}$	18.38%

Table 1.3 presents performance characteristics of the designed solar cell. It is obvious that a solar cell made from material twice as cheap compared to silicon carries much better efficiency.

Notice that the modeled cross section of the device was set to be 1 μm^2, so the current values that are shown on the graph should be converted into current densities to be compared to other solar cells.

1.4.1 Economic Evaluation of CdS/CdSe p-i-i-n Cascaded Solar Cell

The prices for raw materials being compared demonstrate that Si is about twice more expensive than CdS and one and a half times more expensive than CdSe [16, 27, 28]. Thin-film processing, however, depends on technology steps used, i.e., sputtering of Si is still less expensive than sputtering of CdS [29, 30]. In the production of single crystals using epitaxial technology, Si appears to be more expensive [31, 32].

To summarize described above design, the novel heterostructured CdS/CdSe p-i-i-n solar cell was designed to demonstrate the efficiency equal to 18.4% with $V_{oc} = 1.45$ V and short circuit current density $J_{sc} = 23.36$ mA/cm^2, which is better than designed by our group in 2012 solar cell from the same materials [33]. The low cost of thin films [30, 31] of CdS and CdSe compared to Si is a major factor moving technology of II–IV compound semiconductors to the forefront of the solar cells market. The complexity with p-type doping of CdS [34–37] was and still is a subject of studies; however, it would be resolved with selection of epitaxial production technology. We are planning to select epitaxial technology for production of these solar cells in the nearest future. Depending on a selected technology, we will design the buffer layer of graded CdS—CdSe, which should not exceed 30–40 Angstroms in thickness.

1.5   The Number of Intrinsic Regions in Cascaded Solar Cell is Limited

Improved harvesting of solar spectra by cascade of intrinsic segments in a solar cell creates a belief that the number of intrinsic layers in a cascaded solar cell could go infinite. It is true that generation of photo-carriers provides high efficiency solar cells because recombination rate of generated photo-carriers is very small in intrinsic areas. However, requirements for collection of photo-carriers depend on the total thickness of these intrinsic segments.

Thus, the total thickness of intrinsic cascades should not exceed the diffusion length of generated photo-carriers. Below we present the design of a CdS/CdSe p-i-i-i-n solar cell. Optimizing the thickness of each intrinsic segment, we concluded that the total number of intrinsic segments can hardly exceed three layers (see Table 1.4). Efficiency of conventional solar cells proves to be the best when working with the portion of the sun spectrum [38], where energy of photons coming with sunlight is equal to the energy gap of semiconductor material used in the p–n junction of a solar cell. The photons with energy less than the energy gap are not absorbed. However, photons with much higher energy than E_g generate electron–hole pairs, whose scattering heats up the solar element at the same time. These considerations led us to design rules of efficient solar cells step by step. The highly doped but very thin CdS p-layer [39] practically does not absorb coming sunlight. The energy gaps of p-layer and adjacent i-layer are the same, 2.42 eV. This intrinsic layer with energy gap $E_{gi1} = 2.42$ eV is designed to absorb high-energy photons, mostly of green and blue spectra. The range of photon energies E_{ph} absorbed by the second intrinsic layer with energy gap $E_{gi2} = 2.08$ eV is defined by the difference E_{gi1} and E_{gi2}. The energy gap of the third intrinsic layer [40] made of CdSe as well as of the n-doped side of the complete p-i-i-i-n structure is the same, i.e., $E_g = 1.74$ eV (see Fig. 1.10). The energy diagram of all three intrinsic layers is clearly seen. The slope of intrinsic regions demonstrates favorable conditions for collection of photo-electrons and photo-holes. Most of sunlight is absorbed in a cascade of intrinsic layers, where recombination of generated photo carries is zero.

Recently, we attempted to improve the efficiency of solar cells by designing p-i-i-n structures [17], which are efficiently working with two segments of sun spectra. The modeling of current design is described corresponding to the high absorption coefficients of CdSe for the high-energy photons of the solar spectrum and the promising results obtained by our group [17].

Table 1.4 Parameters of potential harvesting of sun spectra by p-i-i-i-n solar cell [18]

Intrinsic layers $CdS_{1-x}Se_x$	Energy band gap (eV)	Est. power available (W m$^{-2)}$	Number of photons cm^{-2} s^{-1}	λ (nm)
(1) (CdS)	2.42	219.96	4.62×10^{16}	512
(2) (CdS$_{0.47}$Se$_{0.53}$)	2.08	148.99	4.1×10^{16}	596
(3) (CdSe)	1.74	176.61	5.86×10^{16}	712

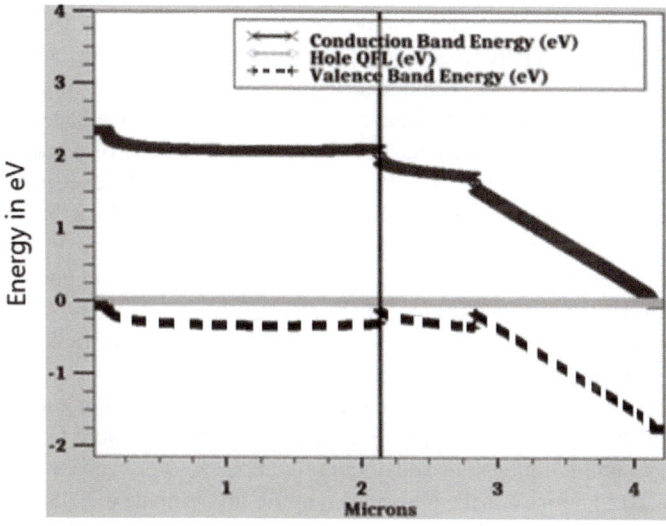

Fig. 1.10 Energy band diagram of the CdS/CdSe cascaded p-i-i-i-n

Figure 1.11 presents the structural profile of the p-i-i-i-n solar cell [18], which consists (region 1) of 0.1-μm-thick p-type CdS top layer with acceptor doping $N_a = 2 \times 10^{18}$ cm^{-3}. The bottom n-layer consists of 0.1-μm-thick CdSe heavily doped $N_d = 1 \times 10^{18}$ cm^{-3}. Sandwiched between the p and n-type layers is a cascade of intrinsic layers (region 2) formed by 2-μm-thick pure CdS ($E_g = 2.42$ eV). This layer is designed to work with the blue part of sun spectra. The second layer designed to work with the green part of spectra (region 3) consists of 0.7-μm-thick CdS$_{0.47}$ Se$_{0.53}$. ($E_g = 2.08$ eV). The 1.3-μm-thick pure CdSe ($E_g = 1.74$ eV) is designed (region 5) to absorb the red part of sun spectra. These intrinsic layers are very lightly

Fig. 1.11 Structure of Cds/CdSe p-i-i-i-n solar cell [18]

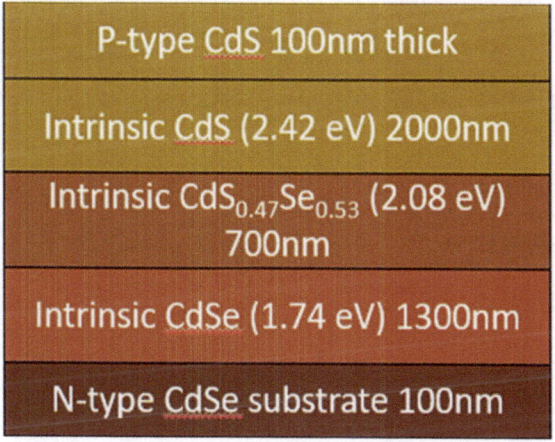

doped with n-type having concentrations of $(4 \times 10^3 \text{ cm}^{-3})$ in CdSe and $CdS_{0.47}$ $Se_{0.53}$ and 10^2 cm^{-3} in pure CdS. The difference in lattice constants of intrinsic CdS and intrinsic CdSe is 3.9%. Manufacturing of our novel design requires detailed analysis of how lattice constants are matching each other in the adjacent layers.

Table 1.4 presents the difference between lattice constants of adjacent intrinsic layers, between layers 1 and 2, i.e., between CdS and $CdS_{0.47}$ $Se_{0.53}$ is greater than 2%. Therefore, to minimize the lattice mismatch, the 2-nm-thick buffer layer of $CdS_{0.84}$ $Se_{0.16}$ with lattice constant 0.586 nm is added there. The mismatch of lattice constants between layers 2 and 3 is only 1.6%, i.e., not significant enough to cause losses due to lattice mismatch (Fig. 1.12).

The results of the finite element analysis have been tabulated in Tables 1.3 and 1.4. The improvement of performance of the solar cell by the addition of the intermediate layer has been assessed through subsequent simulation of another solar cell structure (p-i-i-n) having two 2-μm-thick intrinsic layers made of pure CdS and CdSe, respectively, similar dimensions and doping for anode (p-type) and cathode (n-type) regions. To avoid lattice constant mismatch, buffer layers made of $CdS_{0.84}$ $Se_{0.16}$ and $CdS_{0.47}$ $Se_{0.53}$ are also added at the CdS/CdSe interface. The solar cell characteristics for the p-i-i-i-n are summarized in Table 1.5.

From the plotted and tabulated results, it can rightly be inferred that the p-i-i-i-n structure, with increase in efficiency, fill factor, current density and maximum power,

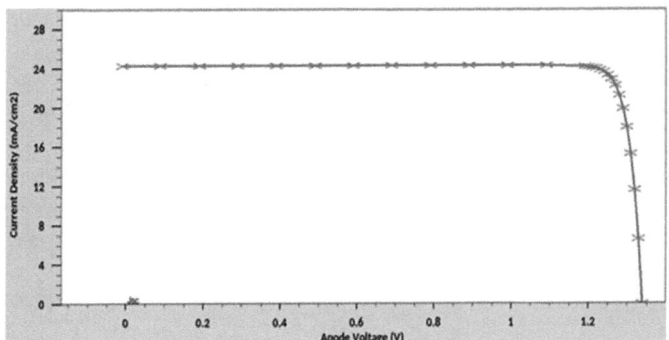

Fig. 1.12 I-V relationship of the CdS/CdSe cascaded p-i-i-i-n solar cell. [18]

Table 1.5 CdS/CdSe p-i-i-i-n solar cell characteristics		
	J_{sc}	24.27 mA/cm^2
	V_{oc}	1.339 V
	P_m	29.3 mW/cm^2
	V_m	1.23 V
	I_m	23.8 mA/cm^2
	FF	90.09%
	E_{ff}	21.2%

provides an efficient alternative to enhancing the overall performance metrics of the solar cell.

In conclusion, we would like to emphasize that the increase in the temperature of our planet has become a significant factor forcing developers of clean energy to speed up the manufacturing of PV resources. Selection of materials is controlled by the cost, and evidently amorphous and crystalline Si will remain dominant materials, although CdS and CdSe are two times cheaper than Si. Next factor to consider is the production cost of efficient solar cells, which is mostly defined by the number of epitaxial layers needed to construct efficient solar cell configuration. Along that line, the number of epitaxial layers in cascaded solar cells is much smaller than the number of layers needed to create tandem solar cells. Readers interested in the specific design of tandem solar cells are advised to look at plenty of publications presented at various photovoltaic conferences. Design of CdS/CdSe heterostructure solar cell with built-in cascade of the three intrinsic layers p-i-i-i-n is in attempt to improve harvesting of solar spectrum. The energy gap of selected intrinsic materials allowed them to absorb red, green and blue portions of sun spectra. The color segments were preselected to carry high flux of photons, on average about 10^{18} cm^{-2}. The absorption coefficients were interpolated from literature sources [41, 42]. Accordingly, thicknesses of intrinsic regions were calculated. It is demonstrated that adding intrinsic layers increases total efficiency of solar cells. However, the number of intrinsic layers in the similar solar converter could not be grown endlessly. Ability to collect all generated by light carriers is limited by the diffusion length of electrons and/or holes. Therefore, the total length of the cascade of intrinsic regions should not exceed the diffusion length of holes. The mismatching of lattice constants in designed heterostructure was evaluated to be not more than 2%. Potential manufacturing of this novel solar cell should use a buffer layer of $CdS_{0.84}Se_{0.16}$. The efficiency of p-i-i-n solar cells in current design is 20.34%. The efficiency of p-i-i-i-n solar cells in current design is 21.2%. Results listed in Tables 1.3 and 1.4 prove the design concept, i.e., careful modeling of intrinsic regions in solar cells, and proper selection of sun spectra segments with high flow of photons leads to increased efficiency of the cell.

References

1. V. Krasnoholovets, Technologies to assist in the energy transition to the century. Invited talk, MRS Webinar on Solar Energy and the Circular Economy, Feb 2021
2. M. Fermeglia, How to avoid the perfect storm: the role of energy and photovoltaics. Invited talk, MRS Webinar on Solar Energy and the Circular Economy, Feb 2021
3. K. Sopian, S.L. Cheow, S.H. Zaidi, An overview of crystalline silicon solar cell technology: past, present, and future, in *AIP Conference Proceedings*, 020004 (2017), p 1877
4. A.W. Kelsey, T.R. Horowitz, B. Smith, A. Ptak, *A Techno-Economic Analysis and Cost Reduction Roadmap for III-V Solar Cells* (National Renewable Energy Laboratory, Golden). NREL/TP-6A20-72103 (2018)
5. M. Tanaka, M. Taguchi, T. Matsuyama, T. Sawada, S. Tsuda, S. Nakano, H. Hanafusa, Y. Kuwano, Development of new a-Si/c-Si heterojunction solar cells: ACJ-HIT (Artificially

constructed junction-heterojunction with intrinsic thin layer). Jpn. J. Appl. Phys. **31**, 3518–3522 (1992)

6. S. Mil'shtein, S. Halilov, Optimization of solar energy harvesting by novel solar cells, in *International Conference on the Physics of Semiconductors*, p. 29, Aug 2014
7. L. Devarakonda, R. Kwende, S. Mil'shtein, Design and optical analysis of corrugated surfaces for silicon solar cell. Scientif. Am. Oct 2012
8. S. Mil'shtein, M. Zinaddinov, Cascaded heterostructured a-Si/c-Si Solar Cell with increased current production, in *Proceedings of IEEE 43rd Internal Conference on Photovoltaic Specialists*, E 45,671, Aug 2016
9. Plotted using an online-tool from http://www.pveducation.org/pvcdrom/design/dlarc
10. Silvaco, Inc., Atlas user's manual. Device Simulation Software", 2021
11. S. De Wolf, A. Descoeudres, Z.C. Holman, C. Ballif, High-efficiency silicon heterojunction solar cells: a review. Green, **2**(1), 7–24 (2012)
12. S. De Wolf, B. Demaurex, A. Descoeudres, C. Ballif, Very fast light-induced degradation of a-Si:H/c-Si (100) interfaces. Phys. Rev. B **83**, 233301 (2011)
13. Z.C. Holman, A. Descoeudres, L. Barraud, F.Z. Fernandez, J.P. Seif et al., Current losses at the front of silicon heterojunction solar cells. IEEE J. Photovolt. **2**(1), 7–15 (2012)
14. Canada Mortage and Housing Corporation, Photovolatic (PV) systems: photovoltaic system overview. http://www.cmhc-schl.gc.ca/en/co/maho/enefcosa/enefcosa_003.cfm.
15. S. Mil'shtein, M. Zinaddinov, N. Tokmoldin and S. Tokmoldin, "Design and fabrication steps of silicon heterostructured p-i-i-n Solar Cell with corrugated Surface", Proceed. IEEE 43rd Intern. Confer. on Photovolt. Specialists, A26, 96, August, 2016.
16. S. Mil'shtein, A. Pillai, S. Sharma, G. Yessayan, Design of cascaded low cost solar cell with CuO substrate, in *International Conference on the Physics of Semiconductors*, ICPS (2012)
17. M. Zinaddinov, S. Mil'shtein, Design of cascaded heterostructured p-i-i-n CdS/CdSe low-cost solar cell, in *Proceedings International IEEE PVSC-44*, 2017
18. S. Mil'shtein, D. Asthana, Solar cells with built-in cascade of intrinsic regions, in *2020 47th IEEE Photovoltaic Specialists Conference (PVSC)*, pp. 1910–1912 (2020)
19. Fraunhofer Institute for Solar Energy Systems, ISE, Photovoltaics report, Nov 2016
20. H. Park, J. Lee, T. Park, S. Lee, W. Yi, Enhancement of photo-current conversion efficiency in a CdS/CdSe quantum-dot-sensitized solar cell incorporated with single-walled carbon nanotubes. J. Nanosci. Nanotechnol. **15**(2), 1614–1617 (2015)
21. X. Mathew, J. Pantoja Enriquez, A. Romeo, A.N. Tiwari, CdTe/CdS solar cells on flexible substrates. Sol. Energy **77**(6), 831–838 (2004)
22. National Renewable Research Laboratory, Cadmium telluride solar cells. Online, Last Accessed: June 2017
23. W.-D. Park, Nanocrystalline CdS thin films prepared by chemical bath deposition, in *Nanotechnology Materials and Devices Conference, 2006. NMDC 2006*, vol. 1 (IEEE, 2006), pp. 460–461, 22–25 Oct 2006
24. S.M. Sze, Physics of Semiconductor Devices, 4th edn. (Wiley, 2006)
25. J.J. Wysoki, P. Rappaport, Effect of temperature on photovoltaic solar energy conversion. J. App. Phys. **31**(3) (1960)
26. S. De Wolf, B. Demaurex, A. Descoeudres, C. Ballif, Very fast light-induced degradation of a-Si:H/c-Si(100) Interfaces. Phys. Rev. B **83**, 233–301 (2011)
27. S.K. Tripathi, A.S. Al-Kabbi, K. Sharma, G.S.S. Saini, Mobility lifetime product in doped and undoped nanocrystalline CdSe. Thin Solid Films, **548**, 406–410 (2013), ISSN 0040-6090. https://doi.org/10.1016/j.tsf.2013.09.008
28. S. Alrich, Silicon. Online, Accessed June 2017. http://www.sigmaaldrich.com/catalog/product/aldrich/267414?lang=en&region=US
29. S. Alrich, Cadmium selenide. Online, Accessed June 2017. http://www.sigmaaldrich.com/catalog/product/aldrich/244600?lang=en&region=US
30. S. Alrich, Cadmium sulfide. Online, Accessed June 2017. http://www.sigmaaldrich.com/catalog/product/aldrich/217921?lang=en&region=US&cm_sp=Insite-_-recent_fixed-_-recent5-2

31. Kurt J. Lesker Company, Cadmium Sulfide (CdS) sputtering targets. Online, Available at: http://www.lesker.com/newweb/deposition_materials/depositionmaterials_sputtertargets_1.cfm?pgid=cd3. Last accessed: June 2017

32. Kurt J. Lesker Company, Silicon (Si) sputtering targets, Online, Accessed June 2017. http://www.lesker.com/newweb/deposition_materials/depositionmaterials_sputtertargets_1.cfm?pgid=si1

33. A.M. Barnett, A. Rothwarf, Thin-film solar cells: unified analysis of their potential. IEEE Trans. Electron Dev. **27**(4), 615–630 (1980)

34. H. Ullal, CdTe PV. Report, National Solar Technology Roadmap (2007)

35. R. Xie, J. Su, M. Li, L. Guo, Structural and photoelectrochemical properties of Cu-doped CdS thin films prepared by ultrasonic spray pyrolysis. Int. J. Photoenergy **2013**, 7 p, Artile ID 620134, 2013

36. K. Poornima, K. Gopala Krishnan, B. Lalitha, M. Raja, CdS quantum dots sensitized Cu doped ZnO nanostructured thin films for solar cell applications. Superlattices Microstruct. **83**, 147–156 (2015)

37. M. Muthusamy, S. Muthukumaran, Effect of Cu-doping on structural, optical and photoluminescence properties of CdS thin films. Optik Int. J. Light Electron Opt. **126**(24), 5200–5206 (2015)

38. https://rredc.nrel.gov/solar//spectra/am1.5/

39. http://www.semiconductors.co.uk/propiivi5410.htm

40. A. Jabbar, S. Abbas, M. Jaduaa, Studying the linear and non-linear optical properties by using z-scan technique for CdS thin films prepared by CBD technique. Appl. Phys. Res. **10**(7) (2018). https://doi.org/10.5539/apr.v10n3p7

41. V.S. Tayn, E.A. Perez-Albuerne, Efficient thin film CdS/CdTe solar cells, in Proceedings of IEEE PVSC **16**, 794 (1982)

42. N.J. Suthan Kissinger, M. Jayachandran, K. Perumal, C. Sanjeevi Raja, Structural and optical properties of electron beam evaporated CdSe. Bull. Mater. Sci. **30**(6), 547–551 (2007)

Chapter 2
Design of Heterostructure Solar Cell Using Non-crystalline a-Si/poly-Si

Abstract History of commercial production of solar cells made from polysilicon ($E_{ff} = 15$–17%) and from amorphous silicon ($E_{ff} = 9$–12%) has accumulated understanding of deficiencies and limitations of these solar cells. With offering simplicity for fabricating poly-Si wafers and ribbons on one hand, these materials suffer from low absorption coefficient. On the other hand, thin-film poly-Si quickly heats up under sun irrigation by high-energy photons. As a result, these solar cells are losing 1% efficiency with every 10 °C increase of the temperature. Very simple technology of making a-Si layers produces materials with high density of dangling bonds, i.e., very high recombination rate of photo-carriers generated by sunlight. Hydrogenation of a-Si films significantly improves performance of these solar cells. However, thin-film a-Si solar cells are effectively blind to part of the sun spectrum with photon energy less than 1.8 eV. In the design of heterostructure solar cells combining a-Si with poly-Si, we were motivated to avoid mentioned above deficiencies of these materials.

2.1 Introduction

The design of solar cells capable of providing best harvesting of solar energy should address the following technical requirements [1–3]:

(a) Ability to harvest energy from widest part of sun spectrum
(b) Offer highest values of absorption coefficient for photons of the selected part of sun spectrum
(c) Ensure highest efficiency of conversion of incident photons into electron–hole pairs or photo-carriers while ensuring lowest recombination rate
(d) Use of corrugated surfaces or ability of tracking solar irradiance increases the number of sun hours during the day.

In addition to these technical parameters, it is important to assess simplicity of manufacturing technology, cost of fabrication of efficient solar cells and longevity of proposed design. The simplicity of fabrication and low cost of mass production directly impact the time needed to replace the footprint of oil and gas energy sources by clean energy. Rapid changes in global climate highlight the limited availability of time.

The heterostructure solar cells containing a sequence of semiconductor materials with different energy gaps provide the best answer to the first requirement. To harvest the most intense flux of photons in the sun spectrum, the top semiconductor layer in heterostructure cells should have $E_g = 1.4$–2.6 eV. Amorphous silicon (a-Si) characterized by $E_g = 1.8$ eV and having high absorption coefficient is one of the best materials for such a top layer.

In the design of our solar cell [4], a thin highly doped layer of a-Si is used. However, high density of dangling bonds, about 10^{20} cm^{-3}, in a-Si [5, 6] causes high recombination rate of photo-carriers. To reduce this high recombination rate, commercial layers of a-Si are subjected to hydrogenation during processing. It is known [7] that plasma hydrogenation carries high processing costs. In the underlying a-Si p-type 0.1-μm-thick layer, the doping gets to the level 10^{18} cm^{-3} to suppress activities of dangling bonds and reduce recombination rate of photo-carriers. The heterojunction is completed by a 100-μm-thick base of poly-Si under the a-Si layers. The detailed design is described in the following section.

2.2 Modeling and Design

The design of the heterojunction solar cell aims to decrease the thickness while optimizing the efficiency of the commercialized polysilicon solar cells [8–10]. Currently available designs of polysilicon solar cells [10–12] require extension of thickness of the n-type base to 180 μm for efficient operation owing to poor values of absorption coefficients depicted in Fig. 2.2 that are offered by the material for the most energetic regions of the solar spectrum as shown in Fig. 2.1.

Fig. 2.1 AM 1.5 solar spectrum

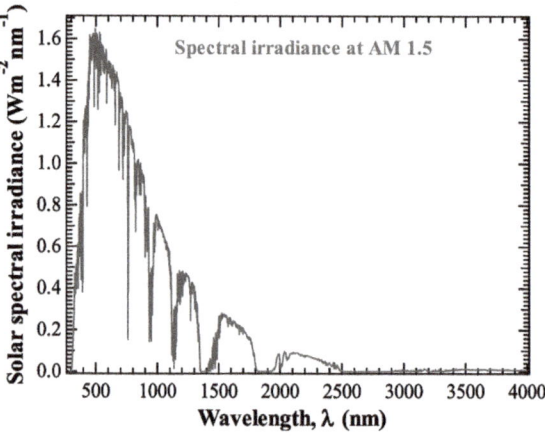

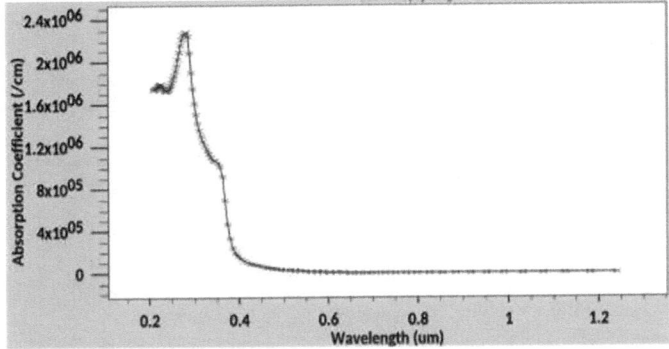

Fig. 2.2 Absorption coefficient of poly-Si for different wavelengths in solar spectrum

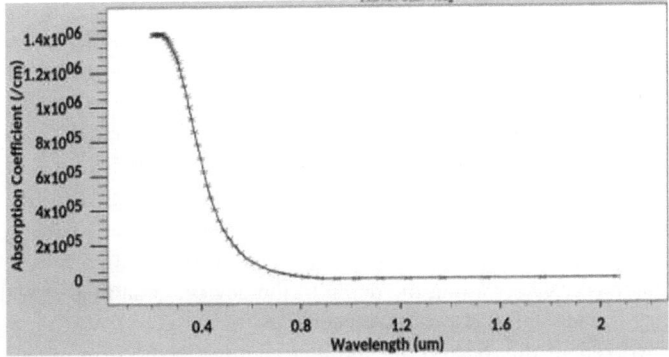

Fig. 2.3 Absorption coefficient of a-Si for different wavelengths in solar spectrum

On the other hand, success of thin-film a-Si p-i-n solar cells is not only realized through high values of absorption coefficients (depicted in Fig. 2.3) for photons lying in the blue, green and red regions of the solar cell, but also high-energy band gap value that helps improve the value for built-in potential and the open circuit voltage of the resulting design while harvesting the most photon abundant region of the solar spectrum.

To decrease fabrication costs and thickness of the commercially available designs, our proposed solution involves growth of a 0.2-μm-thick layer over the surface of a conventional moderately n-type-doped wafer of polysilicon as depicted in Fig. 2.4. Thickness of the n-type wafer has been reduced to 99.1 μm as opposed to 180 μm which is currently popular with existing technology. Growth of this layer can be realized by means of cost-effective growing methods like liquid phase epitaxy (LPE). Doping in the poly-Si wafer has been fixed at 10^{16} donors cm^{-3}, which is common for most solar cells. The bottom-most 0.1-μm-thick section of poly-Si, under the wafer, is heavily doped with donor-type impurity ($N_d = 10^{19}$ cm^{-3}) in order to realize the

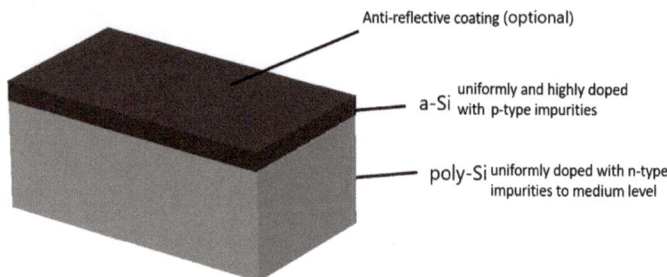

Fig. 2.4 Structure of the solar cell

back contact of the solar cell. The 0.1 μm top section of a-Si is doped with acceptor-type impurities on the order of 10^{19} cm^{-3}, while the 0.1-μm-thick section underneath is p-type doped on the order 10^{18} impurities per cubic centimeter.

The resulting profiles of the energy band diagram at the anode and cathode of the solar cells are depicted in Fig. 2.5a, b, respectively. Value for the overall built-in potential can be approximated at 1 V.

Modeling of this design includes Ohmic contacts at the front and the back contacts of the solar cell. Since it is well evident from a range of publications and experimental works [11–15] that the bulk carrier lifetime associated with the a-Si and polysilicon solar cells is in the order of 200 μs, a similar value has been used for simulating the proposed device. In order to analyze recombination rates associated with the generated carriers along the depth of the device, equations [16] for Auger and Shockley–Read–Hall theories have been used in TCAD SILVACO. Since both materials involved in this design are indirect energy band gap profiles, we have pragmatically excluded radiative recombination theory from the analysis. Therefore, variation in recombination rate throughout the depth of the device is primarily determined by excess carrier injection due to absorption, doping in the material and the availability of the recombination centers or trap states within the forbidden gap of the material.

The density of defect states within the energy band gap of a-Si was studied using the analysis explained in [17, 18] in which they are divided into deep level and tail bands. Each of these bands consists of donor and acceptor-type energy levels.

2.3 Results of Simulation

The complete structure was simulated using finite element analysis to study the response to AM1.5 solar spectrum. High value of fill factor depicted in Table 2.1 indicates optimized performance with decrease in losses due to recombination which is also indicated in the shape of the I/V characteristics which is more rectangular (Fig. 2.6).

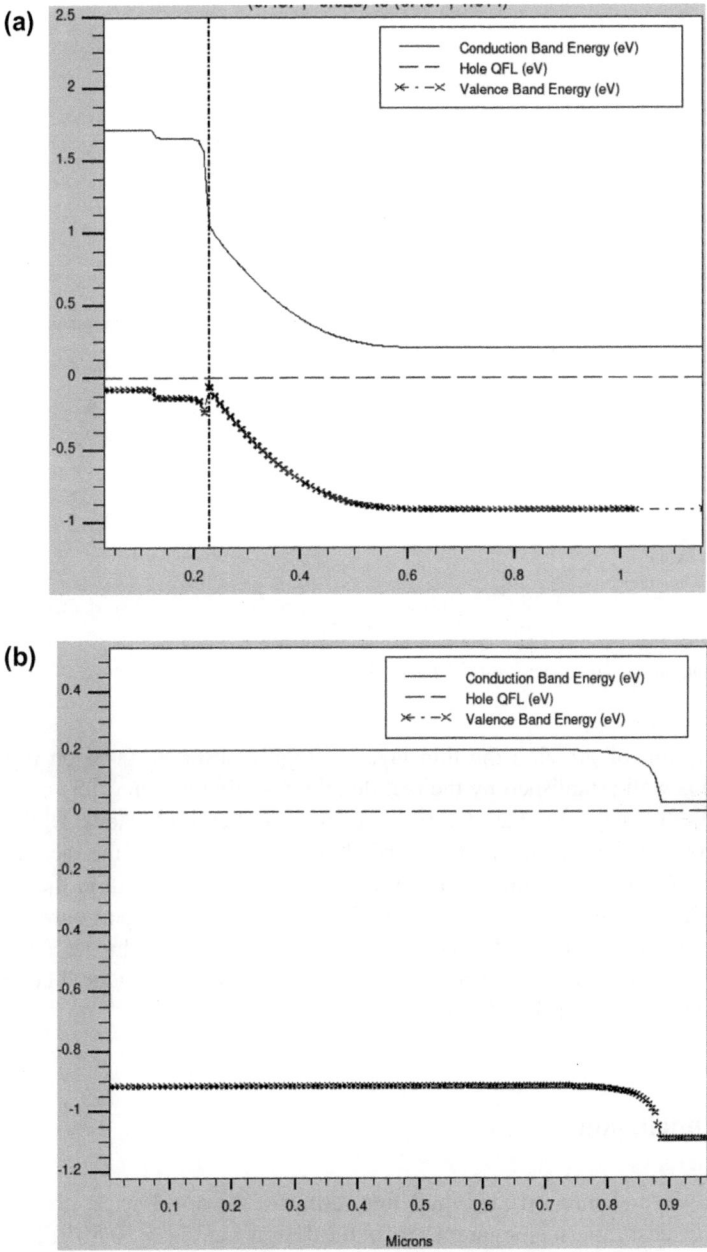

Fig. 2.5 **a** Energy band diagram (y-axis: energy level in eV) at the anode of the solar cell near the anode of the solar cell, **b** Energy band diagram (y-axis: energy level in eV) of the solar cell near the cathode μ (0.95 μm thick section at the bottom) of the solar cell

Table 2.1 Performance characteristics of solar cell

J_{sc}	36.45 mA/cm^2
V_{oc}	0.674 V
P_m	20.94 mW/cm^2
V_m	0.6 V
I_m	34.9 mA/cm^2
FF	85.2%
E_{ff}	20.5%

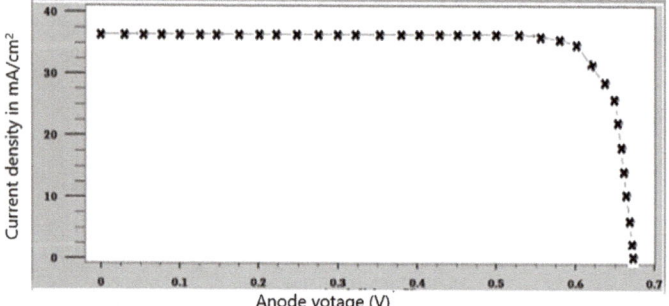

Fig. 2.6 Output current density versus the anode voltage of the solar cell

The benefit of growing the thin layer of highly absorbent a-Si on top of the structure is well established by the fact that the resulting design offers comparable values of efficiency with that of 180-μm-thick polysilicon solar cells. Optimization of thickness of the design has been carried out by ensuring that the recombination rate does not exceed the photogeneration rate at any location within the structure. The efficiency of this skeleton can easily be increased by 2% upon introduction of corrugation and ARC lining that have been discussed in previous chapters. As a result, in terms of performance characteristics, this design can be considered comparable with the state-of-the-art PERC solar cell technology.

2.4 Conclusion

The best way to summarize the modeling results of the novel heterostructure solar cell is to discuss three major parameters of the design, i.e., $E_{ff} = 20.5\%$, FF $= 85.2\%$ and $P_m = 20.94$ mW/cm^2.

Most important is the fact that our design does not carry crystalline materials and therefore is free of complicated considerations about lattice match between adjacent layers. The main specificity of the proposed design is suppression of recombination rates in both amorphous and polycrystalline layers of the structure. Instead of using hydrogenation of dangling bonds, which is costly technology, we propose to use diffusion of impurities in thin amorphous layers and ion implantation of polysilicon.

Design of corrugated surfaces or use of sun tracking systems is optional. There is no preferential etching direction of the amorphous silicon surface. Ability to create corrugated structure of any orientation using beam etching allows to fabricate solar cells with most efficient tracking of sun azimuth angle. Manufacturing of such corrugations can take into consideration the geographical position of the solar cell system.

In the first chapter it was demonstrated that the usage of solar cells for production of electricity is supported by reasonably high ER/EI = 25:1, where EI is the amount of energy invested in installing a generation and ER is the amount of energy produced by it. With the use of low cost of non-crystalline silicon materials in this design and small number of layers that can be grown over the surface of the thin poly-Si wafer through simple Chemical Vapor Deposition (CVD)-based techniques, the ER/EI ratio can significantly be improved.

References

1. T. Ruan, M. Qu, X. Qu, X. Ru, J. Wang, Y. He, K. Zheng, B.H.H. Lin, X. Xu, Y. Zhang, H. Yan, Achieving high efficiency silicon heterojunction solar cells by applying high hydrogen content amorphous silicon as epitaxial-free buffer layers. Thin Solid Films **711**, 138305 (2020)
2. K. Shepard, Z.E. Smith, S. Aljishi, S. Wagner, Kinetics of the generation and annealing of deep defects and recombination centers in amorphous silicon. Appl. Phys. Lett. **53**, 1644–1646 (1988)
3. W. Qarony, M.I. Hossain, M. Khalid Hossain, M. Jalal Uddin, A. Haque, A.R. Saad, Y.H. Tsang, Efficient amorphous silicon solar cells: characterization, optimization, and optical loss analysis. Res. Phys. **7**, 4287–4293 (2017)
4. S. Mil'shtein, D. Asthana, Heterostructure solar cell using non-Crystalline a-Si/poly-Si. US Utility Patent Application No. 17/535, 813
5. Z. Marviab, S. Xu, G. Foroutanb, K. Ostrikovcd, I. Levchenkoad, Plasma-deposited hydrogenated amorphous silicon films: multiscale modelling reveals key processes. RSC Adv. **7**, 19189–19196 (2017)
6. L. Sirleto, G. Coppola, M. Iodice, M. Casalino, M. Gioffrè, I. Rendina, 3-Thermo-optical switches, in *Electronic and Optical Materials, Optical Switches*, ed. by B. Li, S.J. Chua (Woodhead Publishing, 2010), pp. 61–96
7. H. Águas, S. Ram, A. Araújo, D. Gaspar, A. Vicente, S. Filonovich, E. Fortunato, R. Martins, I. Ferreira, Silicon thin film solar cells on commercial tiles. Energy Environ. Sci. **4**, 4620–4632 (2011). https://doi.org/10.1039/c1ee02303a
8. C. Walsh, Solar cell efficiency: n-type v. p-type. CED Greentech (2017). Available at: https://www.cedgreentech.com/article/solar-cell-efficiency-n-type-v-p-type
9. Contributed by: PVEducation.org "Solar cell structure". Available at: https://www.pveducation.org/pvcdrom/solar-cell-operation/solar-cell-structure
10. Contributed by: Aleo Solar "PERC cell technology explained". Available at: https://www.aleo-solar.com/perc-cell-technology-explained/
11. S. Khatavkar, M. Kulasekaran, C.V. Kannan, V. Kumar, K.L. Narsimhan, P.R. Nair, J.M. Vasi, M.A. Contreras, M.F.A.M. van Hest, B.M. Arora, Measurement of relaxation time of excess carriers in Si and CIGS solar cells by modulated electroluminescence technique. Physica Status Solidi. A, Appl. Mater. Sci. **215** (2017)

12. Y. Zeng, Q. Yang, Y. Wan, Z. Yang, M. Liao, Y. Huang, Z. Zhang, X. Guo, Z. Wang, P. Gao, W. Chung-Han, B. Yan, J. Ye, Numerical exploration for structure design and free-energy loss analysis of the high-efficiency polysilicon passivated-contact p-type silicon solar cell. Sol. Energy **178**, 249–256 (2019)
13. B.W.H. van de Loo, B. Macco, M. Schnabel, M.K. Stodolny, A.A. Mewe, D.L. Young, W. Nemeth, P. Stradins, W.M.M. Kessels, On the hydrogenation of Poly-Si passivating contacts by Al_2O_3 and SiN_x thin films. Solar Energy Mater. Solar Cells **215** (2020)
14. Contributed by:pveducation.org "Surface recombination". Available at: https://www.pveduc ation.org/es/fotovoltaica/design-of-silicon-cells/surface-recombination
15. S.W. Glunz et al., Crystalline solar cells-state of the art and future developments (Chap. 1.16), in *Comprehensive Renewable Energy*, vol. 1 (2012)
16. D.A. Neamann, *Semiconductor Physics and Devices* (Chap. 6), 4th edn. (2012), p. 640
17. A.M. Kemp, M. Meunier, C.G. Tannous, Simulations of the amorphous silicon static induction transistor. Solid-State Elect. **32**(2), 149–157 (1989)
18. B.M. Hack, J.G. Shaw, Numerical simulations of amorphous and polycrystalline silicon thin-film transistors, in *Extended Abstracts 22nd International Conference on Solid-State Devices and Materials* (1990), pp. 999–1002

Chapter 3
Enhanced Energy Production by Corrugated Si Solar Sells Installed on Tracking/Anti-tracking Systems

Abstract The first chapter of this study describes modeling and design of the heterostructure cascaded solar cells with potential efficiency of 28–29%. Presence of few intrinsic layers built in a p–n junction, for example, p-i-i-n structure, provides highly efficient conversion of solar energy. Cascaded design could be applied to any heterojunction semiconductor material used in production of solar cells. In the current chapter, we present the design of corrugated surfaces applicable to common solar cells. Corrugation implies design of inverted pyramids on the surface of semiconductor solar cells. It is shown that a corrugated surface increases absorption of solar energy. Careful selection of anti-reflection coating (ARC), proper matching of refractive indices for semiconductor and ARC materials allows to significantly reduce the reflection of solar light. Combination of ARC with corrugated surfaces might be limited by properties of semiconductor materials and/or by the drawbacks associated with the production technology. In recent years, various research groups applied corrugation to surfaces of p-i-n solar cells. However, only researchers of Advanced Electronic Technology Center (AETC) at UMass applied ARC corrugation design to the cascaded Si solar cells. To allow solar cells to work longer hours under some illumination, we offer novel design of tracking/anti-tracking systems. Economic assessment of the system, which combines all three efficient factors, i.e., corrugation, ARC, tracking/ anti-tracking, is discussed at the end of the chapter.

3.1 Introduction

The amount of absorbed energy by planar surface solar cells varies with angle of incidence and reaches maximum for normal incidence. To improve the harvesting of solar energy, the solar panel could be moved by a sun tracking system. Another way of improving absorption with the sun moving around the stationary solar panel was suggested by making inverted pyramids on the surface of silicon solar cells. In the current study, we modeled a corrugated surface for a Si solar cell with two anti-reflection coatings [1]. Our theoretical calculations demonstrate the energy absorption of about 5.1% more than solar cells with inverted pyramid surfaces. For conventional p–n structures, the theoretical harvesting efficiency of a solar cell designed with corrugated surface (case 1) is comparable to efficiency, when sun tracking systems are used (case 2). For both cases, the difference in cost of produced energy (per kWh) is defined by the difference in cost of semiconductor technology and investment into tracking systems. For Si solar cells, the high cost of dual axis tracking system and its maintenance gives an advantage to solar panels with corrugated design. For solar cells made of GaAs and some other semiconductors, the cost ratio might be different. It is helpful to remember that sun tracking increases the number of sun hours daily on average by 30%, and anti-tracking simplifies integration of PV systems into the grid. We discuss the economic impact of tracking/anti-tracking systems on operation of solar farms.

3.2   Design of Corrugation and Anti-reflection Coating

The silicon solar cell using amorphous material, most likely, is the lowest cost semiconductor solar cell in commercial production. However, its efficiency leaves much to be desired. In this section, we strive to reduce the reflection losses by designing appropriate surface texturing and attempting to increase hours of harvesting of solar energy with no sun tracking.

It is known that a planar surface silicon solar cell reflects up to 35% of incident power. That factor prompts the use of anti-reflection coatings (ARC). It has been shown in [1–5] that surface texturing can improve the efficiency of a silicon cell. It was shown that textured surfaces enhance illumination conditions due to internal reflections within the textured surface. Irregular pyramid structure and inverted pyramid described in [2] are commonly used for single silicon solar cells. In this work, texturing of the silicon surface is formatted by a prism-shaped corrugation.

Various types of corrugations are explored and theoretically compared to show the advantage of corrugated versus non-corrugated surfaces, as well as the role of anti-reflection coating and how these can help increase efficiency. Designed corrugation can be produced by selective wet etching methods. Well-known mixture of hydrofluoric acid, nitric acid and acetic acid could be applied as selective etching of (100) surface, securing 1–3 μm etching per second.

Section 3.1 presents the general theory of optical reflection in textured surfaces. Corrugation design is described in Sect. 3.2. Modeling results are discussed in Sect. 3.3.

When the sunlight reaches the surface of a silicon solar cell, about 35% of the light is reflected. By adding an anti-reflection coating of Si_3N_4, this reflection is reduced from 35 to 22% (Fig. 3.1). Surface texturing is another solution [2–4]. Periodic or random texturing of the surface would result in intersurface reflection, i.e., reflection from one surface feature partially illuminates the surrounding surface features for various angles of incidence.

However, the range of angles for which this additional illumination occurs depends on the shape and size of the surface feature and angular position of the sun with respect to the horizon. Adding a single or double films of anti-reflection coating on textured

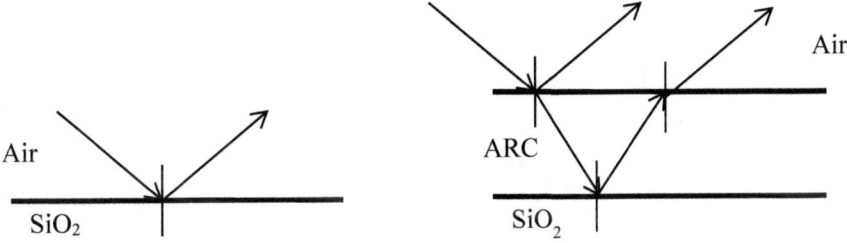

Fig. 3.1 a Surface reflection from Si. 35% of it is reflected. **b** Surface reflection from Si coated with ARC [1]

Fig. 3.2 Light reflection on
a textured surface [1]

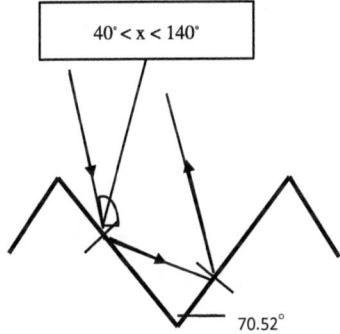

surfaces further reduces reflectance. This does not, however, significantly improve the angle at which additional illumination can occur (Fig. 3.2). Inverted pyramidal structure is the most common type of surface texturing used in silicon solar cells. In current study, we propose a corrugated surface that ensures internal reflection. The effect of corrugation is prominent when the angle of incidence of sun light is between $-54°$ and $54°$ with respect to surface normal.

3.2.1 Design of Corrugation Profile

ARC lined and ARC filled spaces between corrugations on the silicon surface are considered. ARC lined corrugation has ARC lining as shown in Fig. 3.3b while in ARC filled corrugations, the space between corrugations is filled by ARC material as shown in Fig. 3.3a. With this design, there is a limit on the range for full double illumination. It combines the advantage of the planar surface solar cell, which absorbs sunlight from all angles and the double illumination of a corrugated surface to boost absorption of light. A dual layer ARC can further enhances performance of the device.

Table 3.1 presents the percentage losses due to reflectivity for various planar and corrugated surface geometries for normal incidence. It is clear that the presence of

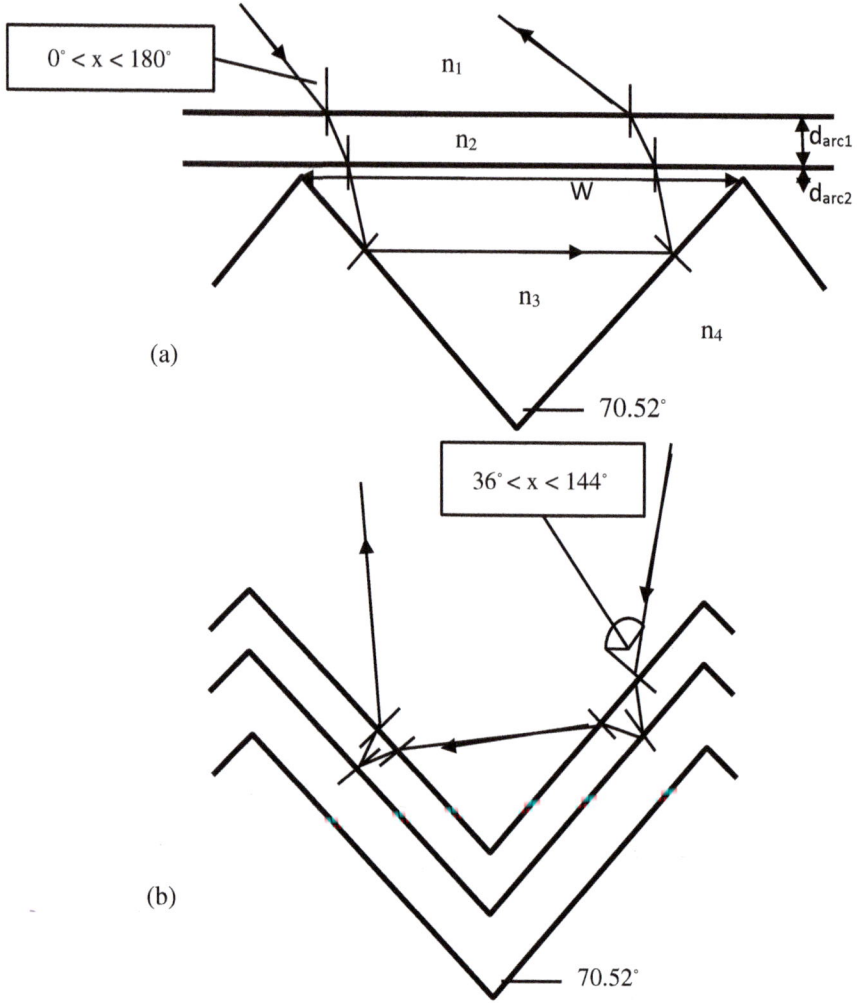

(a)

(b)

Fig. 3.3 **a** Corrugated silicon surface with spaces filled up with ARC. **b** ARC lined corrugation [1]

Table 3.1 Reflection for each surface geometry for normal incidence

Surface geometry	Reflection (%)
Planar without ARC	32.5
Planar with ARC	4
Corrugated without ARC	12.2
Corrugated with thin-film ARC	0.56

corrugation with ARC drastically reduces the reflection losses. Encouraged by these preliminary results, we proceed to analyze the reflection characteristics for a solar cell with ARC lined corrugation and ARC filled corrugations. MATLAB-based ray tracing tools described in [6] were used for our analysis. Absorption in anti-reflection coating is neglected. We compare the surface reflection characters with corrugation to that without corrugation and with two-axial tracking systems.

The total power incident on the cell (global radiation) is a combination of direct, diffused and albedo radiation. All these components depend on the tilt of the panel, geographic location and position of the sun. Hourly variation in direct and diffused solar radiations for a chosen date and location can be obtained from Typical Meteorological Year data (TMY) [7]. Analysis in [8] was used to find the global solar radiation incident at any given time in the day. The solar cell is assumed to be parallel to the surface (no tilt). Hence, albedo radiation can be neglected. Hourly variations in power penetrating the cell on June 4 at Boston, MA, are compared for each of the surface geometries. The total power penetrating the cell can be reduced due to shading for some angles of incidence.

Hence, we compare the density of power penetrating the cell. We speculate that reduction in exposed surface area due to shading is offset by reflections between corrugations. The refractive index of the ARC and silicon are assumed to be real and constant.

Corrugation dimensions: The cheapest way to create corrugation is to selectively etch along a crystallographic surface. Etching a (001) silicon surface will create a <111> oriented surface, and the angle of which with respect to the surface would be about 54.7° [8]. A prism structure with a corrugation angle of 70.6° can be formed with this process. The corrugations width W is arbitrarily chosen to be 20 μm.

ARC filled corrugations: The structure of ARC filling spaces between corrugations is shown in Fig. 3.3a. The thickness of the first ARC layer (d_{arc1}) was chosen to be 10 μm. The thickness of the second ARC layer above the corrugation was chosen to be 10 μm. Overall reflection with respect to angle of incidence is plotted in Fig. 3.4a (from [1]).

Curve (i) represents the fraction reflected for each angle of incidence on a planar surface. Curve (ii) represents the fraction reflected for each angle of incidence on ARC filled corrugated surface.

The density of power (kW/m^2) absorbed by the solar cell for ARC filled corrugated surface (curve ii) is plotted with respect to time of the day for a single day (June 4) for Boston, MA, locations [5] in Fig. 3.4b (from [1]).

The area under the curve gives the total kWh of incident radiation absorbed by the cell. The total kWh absorbed by the cell without corrugation is 6.62 kWh (15 operational hours). The kWh absorbed by the cell with corrugation is 7.1 kWh. There is about 7.6% improvement in the kWh utilized by the cell. This is equivalent to about 1 h increase in the total time of operation of a planar cell.

ARC lined corrugation: Double ARC lined corrugated surface is represented in Fig. 3.3b with another layer of ARC. The thickness of each ARC layer is chosen to be 10 μm. The reflection with respect to angle of incidence is plotted in Fig. 3.5a [1]. Curve (i) represents the fraction reflected for each angle of incidence on a planar

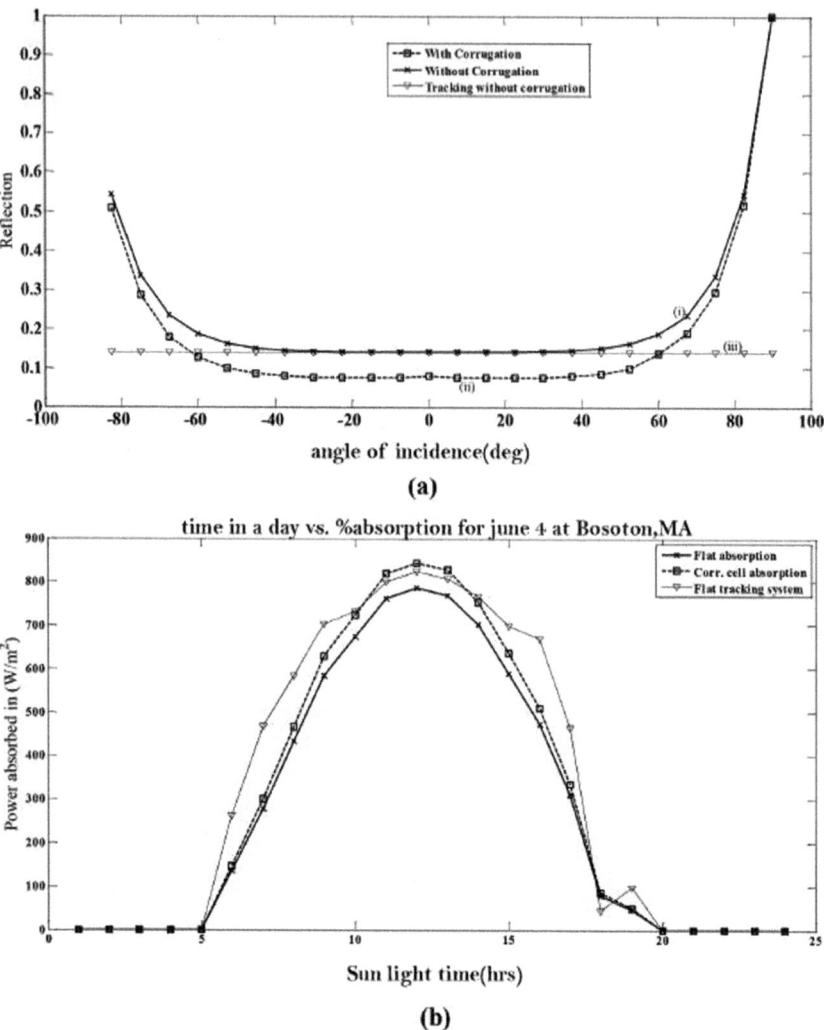

Fig. 3.4 a Fraction reflected versus angle of incidence for planner (i), ARC filled corrugation (ii), planer with tracking (iii). **b** Power penetrating (kW/m²) versus time from sunrise (h) for planer (i), ARC filled corrugation (ii), planer with tracking (iii)

surface. Curve (ii) represents the fraction reflected for each angle of incidence on ARC lined corrugated surface.

The power absorbed by the solar cell for ARC lined corrugated surface (curve ii) is plotted with respect to time of the day for a single day (June 4) for Boston, MA, locations [4] in Fig. 3.5b [3]. From Table 3.2, 7.45 kWh/m² penetrates a solar cell with ARC lined corrugation surface. There is about 12.5% improvement in total

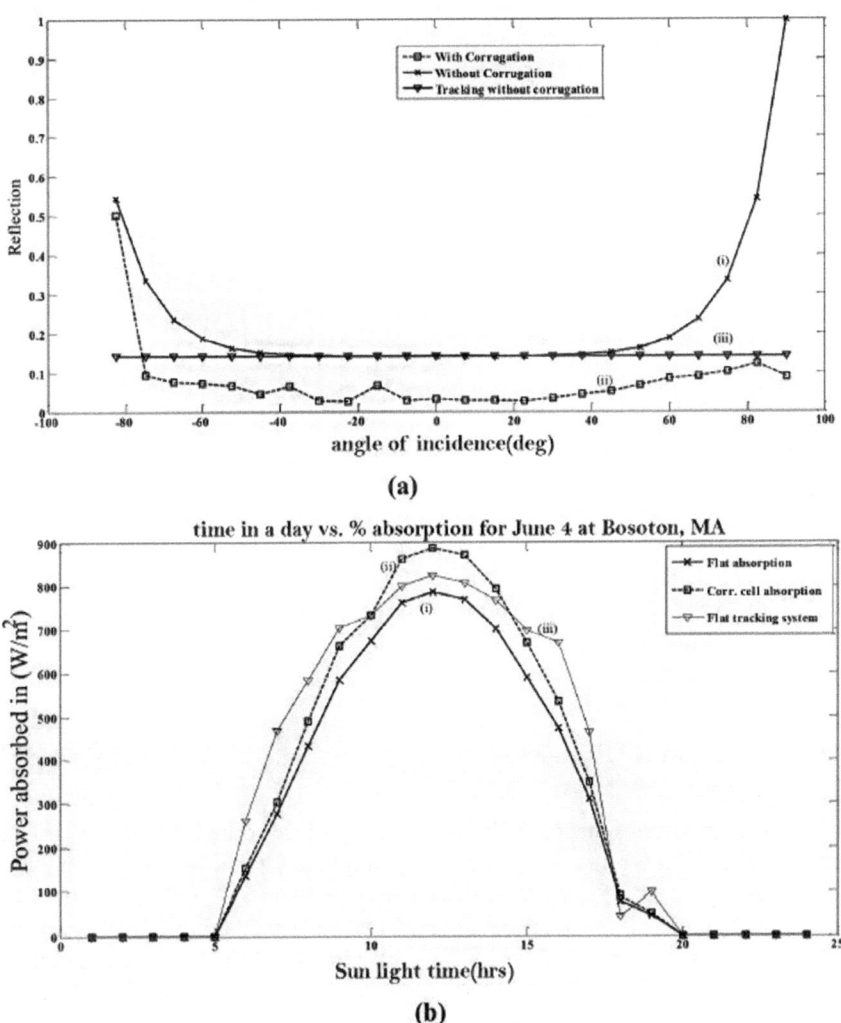

Fig. 3.5 a Fraction reflected versus angle of incidence for planner (i), ARC lined corrugation (ii), planer with tracking (iii). **b** Power penetrating (kW/m²) versus time from sunrise (h) for planer (i), ARC lined corrugation (ii), planer with tracking (iii)

Table 3.2 Power penetrating a cell

	Planar	Corrugated (ARC filled)	Corrugated (ARC lined)	Planar with tracking
kWh/m²	6.62	7.1	7.45	7.9

Table 3.3 Percentage change in total energy (kWh/m^2) absorbed by the cell

	Planar surface (%)	Corrugated (ARC filled spaces) (%)	Corrugated (ARC lined) (%)	Planar with tracking (%)
Planar surface	N/A	+7.6	+12.5	+19.6
Corrugated (ARC filled)	+7.6	N/A	+4.7	−10
Corrugated (ARC lined)	+12.5	+4.7	N/A	−6
Planar with tracking	+19.6	−10	−6	N/A

kWh utilized by the cell. This accounts for about 2 h increase in the total time of operation for a planar solar cell.

Comparison with sun tracking systems: Sun tracking mechanism is used to align the solar cell in the direction of the sun so that the angle of incidence is close to the surface normal. Curve (iii) in Fig. 3.4a (or Fig. 3.5a) represents the reflectance of a cell with a commonly used two-axis tracking system. The hourly kW/m^2 absorbed by a solar cell with a tracking system is given by curve (iii) in Fig. 3.4b (or Fig. 3.5b). From Table 3.3, a total of 7.9 kWh/m^2 is absorbed by a cell when using a two-axis tracking system.

3.3 Results and Discussion

Our modeling indicates that 6.62 kWh/m^2 of energy is absorbed by a planar surface cell with double anti-reflection coating on a single day (June 4) for Boston, MA. Power (kWh/m^2) absorbed by the cell increases to about 7.45 kWh/m^2 with ARC lined corrugation. This is equivalent to about two additional hours of cell operation (12.5%). The energy absorbed is about 7.1 kWh/m^2 for ARC filled corrugation. It is equivalent to about 1 additional hour (7.6%) of cell operation. The kWh absorbed by a double ARC-coated planar surface cell equipped with a tracking system is 7.9 kWh/m^2. This accounts to about 10% improvement over ARC filled corrugation and about 6% improvement over ARC lined corrugations.

Table 3.4 shows percentage improvement in power penetration over planar surface. For normal incidence, ARC filled corrugation behaves similar to inverted pyramid [9] surface textures solar cells. However, performance of ARC lined corrugation exceeds that of inverted pyramid surface [9].

Table 3.4 Percentage improvement over planar cell for normal incidence [1]

Corrugated (ARC filled) (%)	Corrugated (ARC lined) (%)	inverted pyramids [9] (%)
7.23	12.66	7.5

Cost of a c-Si module is about $0.466/kWh ($4/W). The cost of dual axis tracking equipment without inclusion of installation and maintenance expenditure is about 395$/m^2. The total cost of the panels along with the tracking system comes to be about $0.82/kWh ($7/W). In comparison, the cost of the panel with corrugations (assuming 5–7% increase in cost due to selective etching) [10, 11] comes to about $0.489/kWh. The cost benefit would be about 67% over a system with dual axis tracking. In addition, there is a minimum cost associated with repair or maintenance of the tracking systems.

3.4 Summary of Design and Analysis of Different Types of Corrugated Solar Cells

Reflectance characteristic of solar cells with and without corrugation is studied. Power absorbed by cells with a corrugation surface is also compared to that with double ARC-coated planar cells and separately with solar cells equipped with tracking systems. Two types of corrugations surfaces—ARC filled surface and ARC lined corrugation surface—were considered. It is clear that corrugation provides better power absorption than normal planar surface solar cells. When compared to a double ARC-coated planar surface solar cell equipped with dual axis sun tracking, the loss in absorbed energy is only 10% for ARC filled corrugation and only 6% for ARC lined corrugation. ARC lined corrugation seems to perform better than commonly used inverted pyramidal surface texturing. Solar cells equipped with corrugated surfaces are easier to maintain and more economical than planar solar cells equipped with dual axis sun tracking. However, with low costs associated with single axis solar tracking systems as discussed in section 3.10, these systems can economically bolster the harvesting of energy by ARC-lined corrugated solar cells.

3.5 Dual Axis Tracking and Anti-tracking Technology

The major technical challenge that hampers integration of energy systems that harness solar irradiance, into the power grid is the fact that the amount of generated energy varies significantly and erratically during 24 h of daily operation. Appearance of the sun varies with time of the day and weather conditions. Similar issues are encountered by power grid operators when the power output from wind turbines becomes very intermittent during extremely windy conditions as explained in [12, 13]. In response to such circumstances, the technologies for curtailment and boosting of output power, often implemented by means of power electronic converters in commercial and industrial scale solar PV farms, often tend to pose challenges for frequency control of output power. As a result, the output power quality of such generation facilities tends to strain stable operation of the power grid. As of now, current practices of dealing with this challenge include shutting down of wind turbines by system operators during times of extremely windy conditions as reported in [14]. This climatic hurdle has also inhibited further proliferation of wind power generation farms into the energy mix of power grids throughout the world [15, 16].

With limited efficiencies of the order 15% associated with solar PV generation, many technologies have been proposed for optimization. While the idea of concentrated photovoltaics continues to be popular in combination with novel design of solar cells as depicted in [17, 18, 19], large-scale commercial adaptability continues to be a challenge given the high manufacturing costs associated with them. [20, 21] depict the commercial success of single axis tracking technologies that have also been integrated with bifacial PV panels and thermo-photovoltaics. Dual axis tracking technologies depicted in [22, 23] are set to drive the day ahead markets provided economic, once ergonomic and reliable designs of mounting frame and scalable algorithms for their angular position control are realized. However, quantifiable figures affirming the superiority of dual axis tracking systems over single axis trackers still need to be defined.

The current study, which is focused on assessing the eligibility of solar irradiance tracking technologies, in its initial phase, conducts assessment of capacity factors associated with PV farms equipped with single and dual axis tracking systems simulated at different locations on SAM NREL.

In the second part of this work, to decrease the intermittence associated with solar PV, its combination with wind turbines has been proposed. The ultimate goal is to realize a portfolio which is capable of producing energy any day/night as depicted in [24]. The dual axis tracking/anti-tracking solar PV portfolio has been introduced to ensure optimized harvesting of available irradiance. Curtailment of solar generation at times is also mandated due to exhaustion of hosting capacity during times of high congestion in the power grid as explained in [15, 16, 25–32]. This technology is not only proposed to obviate stable operation in terms of output voltage levels and frequency but also to ensure compliance of the power plant with output power limits and curtailments required by system operators.

We have proposed a novel programmable control algorithm compatible with standard automation hardware that allows us to increase or curtail generation from the PV module synchronized with a wind turbine. This section also covers design of a mounting frame for a dual axis tracking system that can achieve variation of output power levels within short time intervals so that factors of intermittence associated with variation of wind speed can be offset. The time interval associated with the design for curtailment of output power has also been evaluated using MATLAB.

Modeling of the overall system is carried out in two stages. In the first stage, a prototype of the mounting frame, motor shafts and the quadrant photodetector was built and tested, an electronic circuit for enabling embedded and synchronized control of the solar panels and the wind turbine for small-scale (800 W) system for dual axis tracking and anti-tracking is proposed, and control algorithms for power optimization and anti-tracking strategy are presented. We extended our idea into the design and modeling of a 1 MW (distributed generation scale) system in the second stage. Real-time operating system (RTOS)-based communication and software architectures have also been proposed for integration of this technology in standardized automation infrastructure. This chapter also provides a brief discussion on the modern design of wind turbines and the various methods of curtailment associated with them.

Our experiments were done in two phases—in the first phase of study, we tested the performance of a scaled down system where control of solar panel position was determined by a conventional dual axis tracking sensor with four photodetectors placed in different quadrants of a circular surface. The second phase includes testing of the synchronizing electronic circuitry that realizes solar tracking and anti-tracking.

3.6 Comparison of Single Axis and Dual Axis Tracking Systems

The basic advantage offered by technologies like solar irradiance tracking is the optimization of direct normal irradiance (DNI) falling onto the surface of the solar panels. The overall generation from solar panels is heavily dependent on the amount of DNI falling on its surface as explained in [33].

Figure 3.6 depicts the global availability of direct normal irradiance over the surface of the earth. In order to find the increase in power production obviated by dual axis tracking systems from single axis tracking configurations at different locations shortlisted in accordance with annual availability of DNI, simulations of 500 kW solar PV plants were carried out in System advisory model (SAM) NREL. Table 3.5 depicts the increase in capacity factor recorded for different locations in the USA.

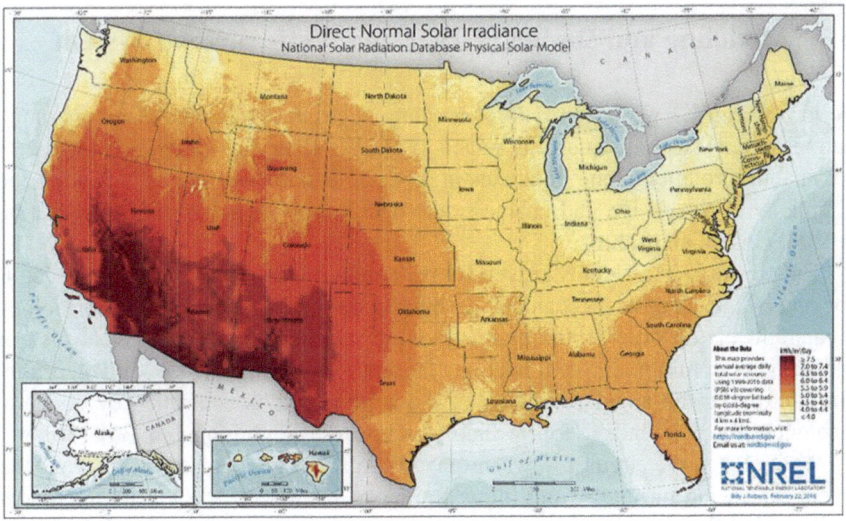

Fig. 3.6 Annual geographical distribution of DNI in the USA (*Source* NREL [16])

Table 3.5 Comparison of capacity factor recorded for dual axis PV portfolio with single axis tracking at different locations

State	Coordinates of location	Increase in capacity factor from single axis module (%)
Massachusetts	42.73° N, −71.1° E	1.8
California	33.61° N, −114.58° E	2.3
North Dakota	46.9° N, −96.8° E	1.9
Arizona	32.13° N, −110.94° E	2.1
California	34.85° N, −116.78° E	2.3
Texas	30.45° N, −99.18° E	1.6
Nevada	40.13° N, -117.98° E	2.3

From the results depicted in Table 3.3, it can well be inferred that the dual axis tracking system depends on the latitude of location of the PV farm and the availability of DNI. The variation of elevation angles of solar irradiance increases above the tropic of cancer which prompts the requirement of dual axis solar tracking, however, decrease in DNI and increase in diffused and scattered radiation at these locations offset the boost in efficiency provided by dual axis tracking. Therefore, best results in terms of gain in efficiency are obtained for latitudes like California, Nevada, North Dakota and Arizona where the availability of solar irradiance is abundant.

The results of this analysis also imply requirement of ergonomic and economic design for dual axis tracking frames that can replace the existing single axis trackers in the market.

3.7 Modeling and Design of Tracking/Anti-tracking System

Following Fig. 3.7, we designed the controlling electronics, selected the motors which can move commercial scale solar panels [34]. It is clear that identical components for enabling rotation along both axes in the solar panel mounting frame add to the simplicity of the design and enable stable operation with pulse width modulation (PWM) control algorithms.

The worm gear assembly driven by the shafts of the DC motors enables self-locking by means of its mechanical mesh when no current is supplied to the motors. Gear ratio of the order 100:1 enables torque multiplication during all cycles of operation with minimal loss in friction. The entire tracking/anti-tracking algorithm for controlling the angular position of the solar panels is controlled by quadrant photodetector as it tends to follow theory and patterns highlighted in [35], [36].

We designed the circuit for synchronized control of the output power from solar panels and wind turbines. Controllers similar to Arduino NANO/UNO microcontroller [37, 38] satisfied our requirements. As shown in Fig. 3.8, the INA240 op amp is used for current sensing with a 0.01 Ω resistor connected across its input terminals.

Fig. 3.7 Prototype of the mounting frame with sensing photodetector [34]

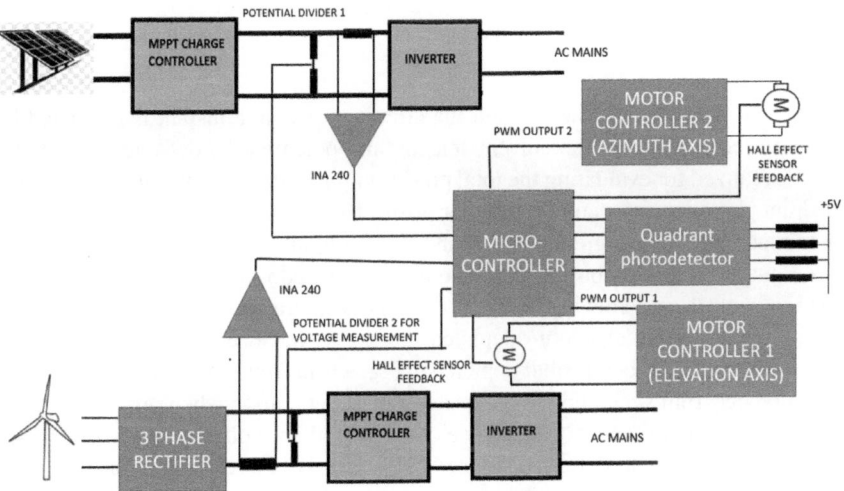

Fig. 3.8 Electronic circuit proposed for enabling synchronized control of output power from solar panels and wind turbine [34]

Since the range of currents flowing within the circuit is expected to vary from 0 to 25 A, the gain of the amplifier is regulated at around 20 V/V.

The fixation of gain enables supply of 0 to 5 V analog input voltage to the Arduino controller. To measure the voltage across the terminals of the combined wind and solar generating unit, the voltage divider circuit supplies another analog input terminal of the Arduino UNO controller after scaling down the input voltage from 24 to 54 V

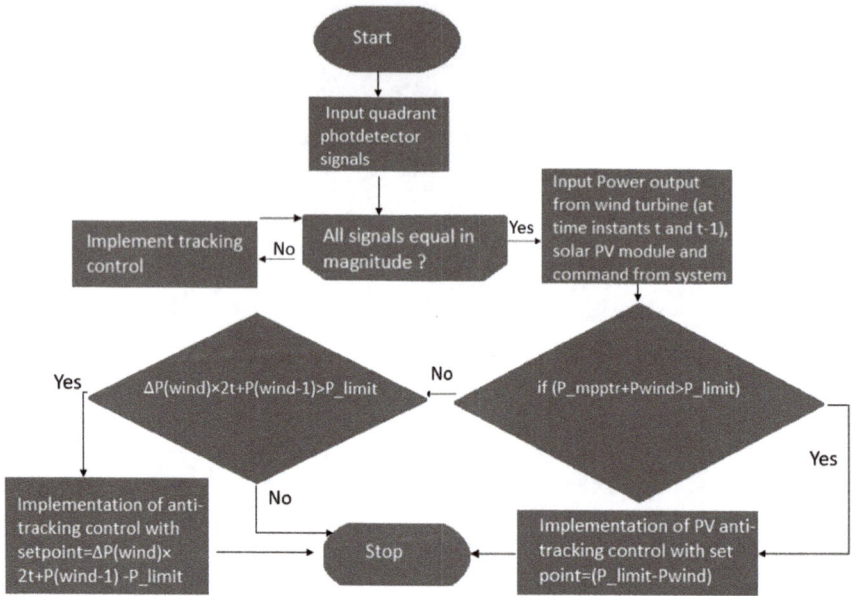

Fig. 3.9 Flowchart depicting the operation of the power curve control algorithm [34]

to 0 to 5 V. The power flow through the circuit at any time instant is evaluated by factoring the gains from the current sensing and potential divider circuits. Similar setup is realized for evaluating the total power output of the system. The output of the quadrant photodetectors and the angular position of the solar panel mounted on it of the miniature system, angular position and power output of the bigger solar panel and power output of the combined system complete the feedback loop of the closed loop position control for achieving accurate results during tracking/anti-tracking cycle.

The programming algorithm depicted in Fig. 3.9 has been proposed for getting embedded in the microcontroller depicted in Fig. 3.9 and meet the goals of optimizing output power from solar and wind, curtailment of generation when prompted by the grid, by changing the angle of incidence of solar irradiance on the panels. The effects of angle of incidence have been described in [39, 40, 41]. Here $P(mpptr)$ and $P(wind)$ are the instantaneous values of the output power obtained from solar PV and wind modules of the setup, respectively. $P(limit)$ is modeled as the threshold finalized in accordance with the curtailment signal from the grid, and $P(wind - 1)$ is the value of output power recorded historically before the lapse of t units of time which is the response time of the hardware and the associated electronics spent in implementing the previous cycle of controls.

The control methodology tends to track solar irradiance only when the total output power generated from wind and solar ($P\_wind + Pmpptr$) is less than $P(limit)$, and the rate of increase of wind power output does not tend to overshoot total generation beyond $P(limit)$ by the end of the current iteration or cycle of control. The first condition ensures compliance with the curtailment signals in case of over production of power from the wind turbine, whereas the second one ensures compliant operation in extremely intermittent weather conditions as far as windy hours are concerned. These conditions also ensure optimized generation of power during times of low generation like the night, no wind, etc. Anti-tracking operation is performed when any of the above conditions are not met. As a result, by means of control strategies, the total output power is limited at P(limit) during cases when wind generation exceeds or tends to overshoot.

3.8 Controller-Based Anti-tracking Algorithm

There is special need for ensuring short time response of the power curtailment mechanism. To ensure quicker response than that offered by the conventional technique which involves iterative cycles of power measurement with incremental changes in angle of tilt from normal, a two-step method has been proposed in this work. The first step of this algorithm approximates the required angle of tilt (θ) for curtailing output solar power down to Pv(limit). Assuming generic equation for output power obtained from PV module, we have

$$P(\text{output}) = V_{oc} \times I_{sc} \times \text{FF} \tag{3.1}$$

where FF is the fill factor, which is dependent on recombination characteristics, ambient temperature and reflection from the upper surface and glass cover of the solar panel. With change in angle of incidence, change in I_{sc} is however significant.

Since intensity of incoming radiation $I(t) \propto E(t)$, where $E(t)$ is the component of the electric field along the normal axis normal to the solar cell. Therefore, for an angle of incidence $= \theta$, the intensity of incoming radiation can be approximated as

$$I(\theta) = I(0) \cos^2(\theta) \tag{3.2}$$

As I_{sc} is directly proportional to the intensity of incoming radiation,

$$I_{sc}(\theta) = I_{sc}(0) \cos^2(\theta) \tag{3.3}$$

After ignoring changes such as minimal variation in open circuit voltage and fill factor (as result of increased reflection losses at oblique angles of incidence), the first step evaluates required angle of tilt (θ) as

$$\theta = \cos^{-1} \sqrt{\left(\frac{P_v(\text{limit})}{P_{\text{mpp}}} \right)} \tag{3.4}$$

where P_{mpp} is the maximum power obtained at normal incidence. First step, as a result, quickly anti-tracks the solar panel to angle of tilt $= \theta$ using PD control. To account for the offset of power caused due to non-consideration of change in fill factor in the first step, the second step initiates an iterative loop to provide final adjustment to the angle of tilt of the solar panel by taking feedback from generated output power from solar panels.

3.9 Modeling of the Up-Scaled System

With increasing proliferation of structured, secure and smart automation technologies in the industrial domain, the requirement of robust, flexible and secure electronic, control and software frameworks has become the driving force behind all developments. Also, the power systems in Europe and the USA, with increasing participation of distributed generation, are set to decentralize and further deregulate as explained in [42] as far as the economics of generation, transmission and load dispatch are concerned. Such developments call for implementation of more advanced control, operation and protective software that can be integrated with electronic and switch gear infrastructure and allow for interoperability between different modules and microgrids. In order to address these concerns, the International Electrotechnical Communication (IEC) has proposed standard architectures and protocols like IEC 61850 and IEC 61970 to obviate interoperable communication between different modules of the automation network as explained in [43]. Reference [44], however, also discusses the adoption of IEC 61499 for smart grid architecture. The modules comprise logical nodes or intelligent electronic devices (IEDs) like programmable logic controllers (PLCs) which are microprocessor-based systems with embedded intelligence, control the actuators, securely exchange data and therefore implement complex automation algorithms. The logical nodes comprise controllers of various power electronic drives of PV farms, shunt reactors in substations, etc.

Materialization of the up-scaled version of the proposed algorithm, therefore, requires the associated infrastructure power meters, automation framework and actuation modules to be compliant with these protocols. Power meters compliant with these standards are already manufactured by companies like Siemens, ION, etc.

Since power system automation is centered around the use of real-time operating systems which are suited for deadline-based service and error mitigation, operating systems like freeRTOS and QNX which are also compliant can be easily incorporated. Stepper motor drives manufactured by most of the companies like Siemens and Schneider Electric are also compliant with these standards. The prospective design of the hybrid wind and PV power plant with wind and solar tracking and anti-tracking can therefore be designed by means of compliant power meters operating systems and actuators. As shown in Fig. 3.10, the output of the power meters and photodetectors can be used to model error conditions for the operating systems in accordance with the flowchart depicted in Fig. 3.9. The intelligence of the RTOS will then convert the conditional algorithm into a priority-based scheduling problem and direct actuators of our system comprising solar tracking/anti-tracking, pitch angle control for tracking/anti-tracking of wind turbines while ensuring desirable operation

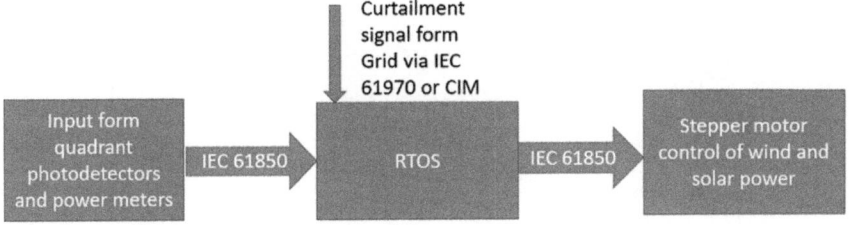

Fig. 3.10 Communication scheme for exchange of information for implementation of tracking/anti-tracking [34]

of the plant. The real-time operating system architecture can be used to realize the implementation of tracking/anti-tracking algorithms. The proposed architecture has been derived from the fact that our tracking/anti-tracking algorithm changes functionality (switches from tracking to anti-tracking and vice versa) upon reception of curtailment signals from the grid and power signals from the wattmeter during highly intermittent weather conditions.

The operation of the prototype of the mounting frame depicted in Fig. 3.7 was successfully tested for solar tracking and self-locking. With standard 12 V DC supply being fed to the motors, the motors are able to track the angle of irradiance within a time interval of 0.1 s. The performance assessment of the quick response anti-tracking algorithm was carried out in the MATLAB/Simulink environment.

The model consisted of a stepper motor with torque characteristics comparable with the motors used in Fig. 3.7, a set of two solar panels with combined power generation of the order 312 W. A set of microcontroller codes for realizing anti-tracking were also written in a connected module. The angular displacement of the stepper motor was related to change in irradiance in compliance with Eq. 3.1, which in turn served as input to the solar cell.

In order to ensure stable operation and ensure limited consumption of electrical power, the angular speed of the motor was limited to approximately $\pi/5$ rad/s. The first stage of the anti-tracking algorithm involved tracking of θ (angular position) using PD control for angular speed of the motor. Favorable responses were obtained for K_p (coefficient proportional to error) $= 2$ s^{-1} and K_d (coefficient proportional to derivative of error) $= 0.05$. The first stage of the anti-tracking algorithm was timed for 3.5 s and corresponds to the minimum in Fig. 3.11 (shown in red), which depicts the threshold power level let by the curtailment signal in blue.

The second stage which involves decremental and cyclic changes of two steps in the angular position of the motor follows stage 1. From Table 3.6, the average response time for different levels of curtailment can be approximated at 5.02 s.

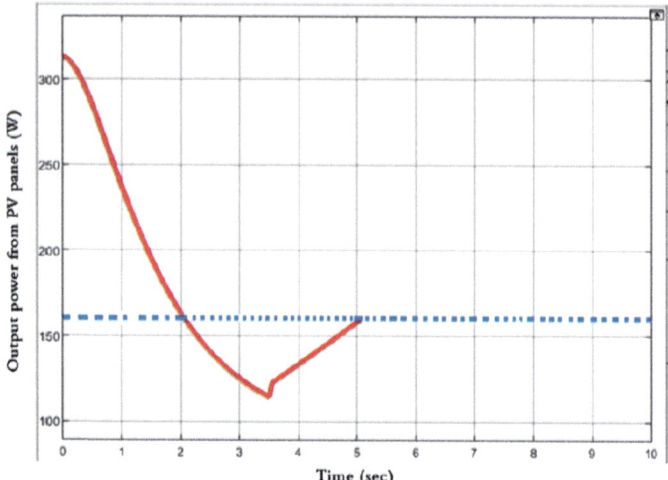

Fig. 3.11 Performance of the anti-tracking algorithm (Dark line: output power, dashed line: set point for power limit)

Table 3.6 Response time for achievement of various levels of curtailment from optimum value of 312	Level of curtailment (Fraction of optimum power)	Time response (s)
	0.25	5.5
	0.33	5.5
	0.5	5.0
	0.67	4.8
	0.75	4.6

3.10 Combination of ARC Lined Solar Panels with Single Axis Tracking

As vividly described in Fig. 3.4, the fundamental leverage offered by corrugated solar cells lined with ARC is minimization of losses due to reflection at any angle of incidence. On the other hand, the success of use tracking technology in increasing the effective duration of PV power generation is based on the philosophy of optimizing the angle of incidence for ensuring maximized power output of panels. While the economic feasibility of ARC lined corrugated solar cell fabrication technology has already been confirmed in the previous sections, the popularity of tracking technologies in recent years has allowed rapid downfall in the capital and operations and maintenance costs associated with the latter. Therefore, a combination of the two systems, which, not only ensures maximized generation by ensuring reduction of reflection losses but offers possibility of economic operation also. As a result, at any time instant during the day, solar panels can be oriented at angles offering maximum power complemented with optimized reflection losses.

From Figs. 3.4 and 3.5, it can also well be inferred that reflection losses in ARC lines corrugated cells begin to dominate once the angle of incidence increases beyond 60°. The associated span of 120° can easily cover requirements of ensuring maximum absorption as far as solar elevation angles at all geographic locations are concerned. Solar energy convertors and windmill generators are well-established technologies. As a result, the ARC corrugated solar cells can eliminate the requirement of tracking systems for solar elevation angles while further increasing the figures of additional power generation enumerated in section II by 0.1%.

3.11   Economic Assessment of Combination of ARC Lined Solar Panels with Single Axis Tracking

Prospects for commercial adoption of this combination are entirely dependent on the evaluated figure of levelized cost of energy (LCOE) [43–48]. Since capital costs, capacity factor, operations and maintenance costs and life of economic operation are the key technical determinants of economic feasibility of any power generation system, the subsections that follow evaluate the impact of this combination on these metrics. Eventually, the value of LCOE is calculated for 1 MW scale generation capacity.

(a)   **Capital cost**
Corrugation of PV panels increases the capital costs associated with manufacturing of PV panels by 12.66% as discussed in previous sections.
From the recent report summarizing the PV system costs in the USA for the first quarter of 2020 [30], it can well be inferred that approximately 45% of the total capital costs associated with solar PV farms is accounted for by PV panels. Moreover, as a result of approximation of data presented in [49], the overall increase in capital costs for the entire system can be pitched at 5.7%. Therefore, as shown in Table 3.8, the capital costs associated with the system are pitched at $1.5 million.

(b)   **Operations and Maintenance Costs**
The data for operations and maintenance costs associated with single axis tracking systems is summarized in Table 3.7. Introduction of ARC lined corrugated panels into the systems adds negligible liability to budget for operations and maintenance since allocation for module cleaning suffices all requirements.

(c)   **Lifetime of Operation**
The technology of anti-tracking along the azimuth axis has been introduced as a better alternative to conventional methods of curtailment that are currently implemented by the inverter for existing PV plants. Curtailment by electronic means often takes a huge toll on the lifetime of inverters. Since PV inverter repairs account for 50% of unscheduled operation and maintenance costs, a voluminous fraction is saved in maintenance of damaged IGBTs and melting contacts due to heat generated by dissipated power within the overall electronic

Table 3.7 Comparative breakdown of operations and maintenance costs in USD/kWh for varying scales of generation capacity and portfolio (fixed tilt and single axis tracking systems) [30]

Service	Residential	Commercial	(Utility scale) fixed-tilt	(Utility scale) tracking
Module cleaning and vegetation management	$0.80	$2.70	$3.30	$3.30
System inspection and monitoring	$2.72	$4.97	$1.79	$2.43
Component parts replacement	$4.55	$0.93	$0.55	$0.87
Module replacement	$0.82	$0.82	$0.91	$0.91
Inverter replacement	$10.0	$5.54	$3.77	$3.77
Operations administration	$2.60	$2.57	$2.50	$2.86

Table 3.8 Impact of ARC lined corrugated solar panels mounted on single axis tracking frames

Metric	Value
Capacity factor	26.16%
Operations and maintenance costs	$10,000
Capital costs	$1.5 million
Life of economic operation	51 years
Levelized cost of energy	3.1cents/kWh

circuitry during hours of curtailment and periods of very low generation. As a result, in Table 3.8, the increase in operations and maintenance cost for dual axis PV tracking systems gets offset by a decrease of 15% to register approximate savings of 35%. This technology which changes incident angle for solar irradiance on the PV panels as a means of curtailment prevents damage to all the electronic components of the system. The implementation of this technology requires modification in the embedded system code of the automation software and therefore incurs negligible financing in favor of capital and operations and maintenance costs. PV inverters generally account for 20% of the plant construction costs of a PV farm; therefore, in agreement with the assessment conducted in [26, 30, 50, 51] for the impact of heat dissipation on the lifetime of PV inverters, an increase of 54% in the life of economic operation has been considered for evaluation of LCOE. From the analysis conducted in [35], it is clear that by keeping the angular inclination of the solar panels fixed at certain with respect to the azimuth and the elevation axes, owing to decrease of 0.03% in degradation rate, their lifetime can be increased significantly. By assuming nominal life of operation for solar panels as 30 years, and factoring in this value of decrease in degradation rate which can be realized through anti-tracking, the overall useful life of operation can be increased up to 51 years as shown in table 3.8.

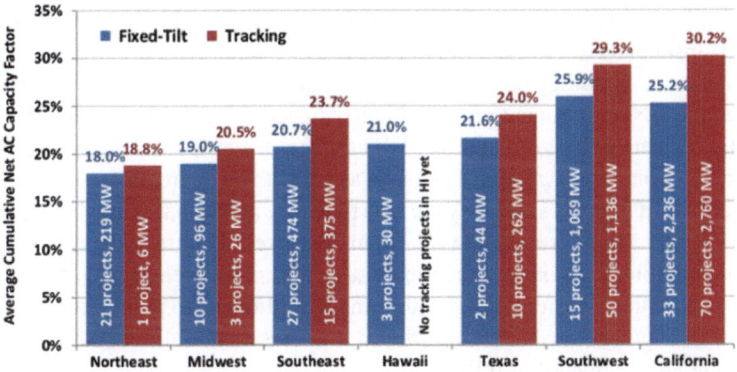

Fig. 3.12 Improvement in capacity factor offered by single axis tracking systems throughout the USA (from [52])

(d) **Capacity factor**

Figure 3.12 shows the data for capacity factor of the single axis tracking port-folios installed throughout the US sourced from [52]. From the depicted data and analysis of [49], the average capacity factor associated with single axis tracking solar PV portfolios located throughout the USA can be assumed as 24%.

This figure is further bolstered by the ARC lines corrugated solar panels that not only compensate for dual axis tracking by increasing the capacity factor 2.04% but also ensure 0.12% gain in output generation (extrapolated from Fig. 3.4). As a result, the average capacity factor used for analysis is 26.16%.

The formula used for evaluation of LCOE can be written as

$$L_{COE} = \frac{\sum_{t=1}^{t=n} \frac{I_t + M_t + F_t}{(1+r)^t}}{\sum_{t=1}^{t=n} \frac{E_t}{(1+r)^t}} \tag{3.5}$$

where I_t is investment expenditure in year t which covers investment in capital costs before the lapse of lead time interval after which the EPC contract ends and the facility begins to operate, M_t is the operations and maintenance cost, F_t represents fuel costs, E_t represents energy generation for the year t, r is the discount rate implying depreciation, and n is the total life of operation.

3.12 Conclusion

In this chapter, comparison between dual axis and single axis tracking systems is based on results of an energy harvesting model that accounts for direct normal irra-diance (DNI), diffused horizontal irradiance (DHI) and global horizontal irradiance

(GHI) while simulating total energy yield from small/distributed generation (DG) scale PV installations. However, the tracking systems discussed in this chapter are based on the methodology of tracking sun (TS) position which is known to primarily optimize the DNI component of available irradiance. It has been established in [53] that solar trackers of best orientation (TBO systems), which aim to optimize capture diffused component of available irradiance along with DNI, are more relevant at latitudes northward of $60°N$ since they are able to realize additional energy yield of more than 1.8% over a meteorological year when compared with tracking systems based on TS methodology. Main implication of such findings and the scale of generation capacity taken into consideration for this analysis allow us to pragmatically neglect concerns of angular distribution of diffused irradiance since all locations covered in the chapter are southward of $60°$ N.

Analysis of economic feasibility can be summarized through increase in economic life of operation leveraged by anti-tracking technology; decrease in capital, operations and maintenance costs enabled by ARC lined corrugated solar cells and single axis tracking makes this portfolio an eligible candidate for adoption at the commercial scale. Large-scale proliferation of single axis tracking systems and possibility of cost-effective customization of silicon photovoltaic fabrication also make this combination economically feasible. It is also well understood that the value of LCOE evaluated for this combination is well below the 4.3 cents/kWh which is currently the average figure for LCOE associated with solar PV portfolios.

References

1. L. Devarakonda, R. Kwende, S. Mil'shtein, Design and optical analysis of corrugated surfaces for silicon solar cell. Scientif. Am. (2012)
2. B. Dale, H.G. Rudenberg, High efficiency silicon solar cells, in *Proceedings of the 14th Annual Power Sources Conference* (U.S. Army Signal Research and Development Lab, 1960), p. 22
3. W.L. Bailey, M.G. Coleman, C.B. Harris, I.A. Lesk, United States Patent: 4137123—Texture etching of silicon: method (1979)
4. J. Zhao, A. Wang, M.A. Green, F. Ferrazza, 19.8% efficient "honeycomb" textured multicrystalline and 24.4% monocrystalline silicon solar cells. Appl. Phys. Lett. **73**(4), 1991–1993 (1998)
5. S.R. Wenham, M.A. Green, Buried contact solar cell, Patent No. 4726850, Feb 1988
6. Available at: http://www.mysimlabs.com/. Last Accessed: July 2020
7. Available at: http://rredc.nrel.gov/solar/old_data/nsrdb/1991-2005/tmy3/. Last Accessed July 2020
8. Available at: http://pveducation.org/pvcdrom/properties-of-sunlight/calculation-fo-solar-ins olation. Last Accessed July 2021
9. A.W. Smith, Analysis of textured solar cells at various angles of incidence: Fresnel concentration to 500 suns. Solar Energy Mater. Solar Cells. **32**(1), 37–51 (1994)
10. Available at: http://en.wikipedia.org/wiki/Etching_%28microfabrication%29. Last Accessed: July 2020
11. Available at; http://americansolareconomy.blogspot.com/2009/03/does-solar-tracking-make-sense.html. Last Accessed: July 2020

12. H. Li, Z. Lu, Y. Qiao, W. Wang, Risk assessment of power system with high penetration of wind power considering negative peak shaving and extreme weather conditions, in *2014 IEEE PES General Meeting\Conference & Exposition*, pp. 1–5 (2014)
13. 30m bill for wind turbines that don't work when it's windy. *Daily Mail* [London, England], 26 Dec 2013, p. 16
14. J.F. Martínez, M. Steiner, M. Wiesenfarth, G. Siefer, S.W. Glunz, F. Dimroth, Hybrid bifacial CPV power output beyond 350 W/m^2, in *International 47th IEEE PVSC Conference Proceedings* (2020)
15. Available at: https://www.nrel.gov/gis/wind.html. Last Accessed: July 2020
16. N. Kelley, M. Shirazi, D. Jager, S. Wilde, J. Adams, M. Buhl, P. Sullivan, E. Patton Lamar low-level jet project interim report, NREL/TP-500-34593 (2004)
17. Y.O. Mayon, M. Stocks, K. Booker, C. Jones, A. Bakers, GaAs/Silicon tandem microconcentrator module, in *IEEE 47 PVSC Conference Proceedings* (2020)
18. C. Zhang, E. Armour, R. King, C. Honsberg, Low concentrated photovoltaics for III-V solar cells, in *2018 IEEE 7th World Conference on Photovoltaic Energy Conversion (WCPEC) (A Joint Conference of 45th IEEE PVSC, 28th PVSEC & 34th EU PVSEC)* (2018), pp. 0975–0979
19. K.R. McIntosh, M.D. Abbott, B.A. Sudbury, J. Nagyvary, K. Lee, L. Creasy, J. Sharp, J. Crimmins, D. Zirzow, Simulation and measurement of monofacial and bifacial modules in a 1D tracking system, in *IEEE 47 PVSC Conference Proceedings* (2020)
20. K.R. McIntosh, M.D. Abbott, B.A. Sudbury, C. Zak, G. Loomis, A. Mayer, J. Meydbrey, Irradiance on the upper and lower modules of a two-high bifacial tracking system, in *IEEE 47 PVSC Conference Proceedings* (2020)
21. Y. Yao, H. Yeguang, S. Gao, G. Yang, D. Jinguang, A multipurpose dual-axis solar tracker with two tracking strategies. Renewable Energy **72**, 88–98 (2014)
22. H. Fathabadi, Comparative study between two novel sensorless and sensor based dual-axis solar trackers. Sol. Energy **138**, 67–76 (2016)
23. T. Tao, H. Zheng, Y. Su, S.B. Riffat, A novel combined solar concentration/wind augmentation system: constructions and preliminary testing of a prototype. Appl. Thermal Eng. **31**(17–18), 3664–3668 (2011)
24. S. Kahrobahee, V. Mhr, Probabilistic analysis of PV curtailment impact on distribution circuit hosting capacity, in *IEEE 47 PVSC Conference Proceedings* (2020)
25. H. Böök, A. Poikonen, A. Aarva, T. Mielonen, M.R.A. Pitkänen, A.V. Lindfors, Photovoltaic system modeling: a validation study at high latitudes with implementation of a novel DNI quality control method. Solar Energy **204**, 316–329 (2020)
26. A. Sangwongwanich, Y. Yang, D. Sera, F. Blaabjerg, Lifetime evaluation of grid-connected PV inverters considering panel degradation rates and installation sites. IEEE Trans. Power Electron. 1–10 (2017)
27. G. Saikrishna, S.K. Parida, R.K. Behera, Effect of parasitic resistance in solar photovoltaic panel under partial shaded condition, in *International Conference on Energy Systems and Applications* (2015), pp. 396–401
28. F.R.S. Sevillaa, D. Parrab, N. Wyrschc M.K. Patel, F. Kienzle, P. Korba, Techno-economic analysis of battery storage and curtailment in a distribution grid with high PV penetration. J. Energy Storage **17**, 73–83 (2018)
29. L. Bird, J. Cochran, X. Wang, Wind and solar energy curtailment: experience and practices in the United States. Technical Report NREL/TP-6A20-60983 (2014)
30. R. Wiser, M. Bolinger, J. Seel, *Benchmarking Utility-Scale PV Operational Expenses and Project Lifetimes: Results from a Survey of U.S. Solar Industry Professionals* (Lawrence Berkeley National Laboratory, June 2020)
31. R. Fu, D. Feldman, R. Margolis, *U.S. Solar Photovoltaic System Cost Benchmark: Q1 2018* (National Renewable Energy Laboratory, 2020)
32. M. Zinaddinov, S. Mil'shtein, Solar tracking with anti-tracking support for ancillary service, in *Proceed International IEEE PVSC 46 Conference* (2019)
33. M. Zinaddinov, S. Mil'shtein, Solar tracking with anti-tracking support for ancillary service, in *International 46th IEEE PVSC Conference Proceedings* (2019)

34. S. Mil'shtein, D. Asthana, J. Scheminger, S. Hummer, Energy production by solar tracking/Anti-tracking combined with wind turbine, in *2020 47th IEEE Photovoltaic Specialists Conference (PVSC)*, pp. 1635–1639 (2020)
35. A.A. Glick, N. Bossuyt, J. et al. Utility-scale solar PV performance enhancements through system-level modifications. Sci. Rep. 10, 10505 (2020). https://doi.org/10.1038/s41598-020-66347-5
36. Available at: https://www.teachengineering.org/lessons/view/cub_pveff_lesson01. Last Accessed: June 2020
37. J.F. Lee, N.A. Rahim, Performance comparison of dual-axis solar tracker vs static solar system in Malaysia, in *2013 IEEE Conference on Clean Energy and Technology (CEAT), Lankgkawi* (2013), pp. 102–107. https://doi.org/10.1109/CEAT.2013.6775608
38. T. Kaur, S. Mahajan, S. Verma, Priyanka, J. Gambhir, Arduino based low cost active dual axis solar tracker, in *2016 IEEE 1st International Conference on Power Electronics, Intelligent Control and Energy Systems (ICPEICES), Delhi* (2016), pp. 1–5
39. J.-A. Jiang, J.-C. Wang, K.-C. Kuo, Y.-L. Su, J.-C. Shieh, On evaluating the effects of the incident angle on the energy harvesting performance and MPP estimation of PV modules. Int. J. Energy Res. **38**, 1304–1317 (2014)
40. A. Wilson, R. Ross, Angle-of-incidence effects on module power and energy performance, in *Progress Report 21 and Proceedings of the 21st Project Integration Meeting Jet Propulsion Laboratory, Pasadena, CA* (1983), pp. 423–426
41. O. Zinaman, M. Miller, A. Adil, D. Arent, J. Cochran, R. Vora, S. Aggarwal, M. Bipath, C. Linvill, A. David, M. Futch, E.V. Arcos, J.M. Valenzuela, E. Martinot, M. Bazilian, K. Pillai, Power systems of the future. NREL/TP-6A20-62611, Feb 2015
42. N. Etherden, V. Vyatkin, M.H.J. Bollen, Virtual power plant for grid services using IEC 61850. IEEE Trans. Industr. Inf. **12**(1), 437–447 (2016). https://doi.org/10.1109/TII.2015.2414354
43. G. Saeid, M.S. Davood, M. Behrooz, M. Mir, A. Mehrabian, B, Razeghi, S.S. Sebtahmadi, IEC61850 implementation in localization of power substation automation system (2014). https://doi.org/10.1109/ROMA.2014.7295868
44. Available at: https://www.nrel.gov/analysis/tech-cap-factor.html. Last Accessed: July 2020
45. N.P. Kumar, K. Balaraman, C.S.R. Atla, Optimizing system elements for hybrid wind–solar PV power plant, in *Biennial International Conference on Power and Energy Systems: Towards Sustainable Energy (PESTSE)* (2016), pp. 1–6
46. W. Choia, M.B. Patea, R.D. Warrenb, R.M. Nelsonc, An economic analysis comparison of stationary and dual-axis tracking grid-connected photovoltaic systems in the US Upper Midwest. Int. J. Sustain. Energy **37**(5), 455–478 (2018)
47. N. Nandasiri, C. Pang, V. Aravinthan, Marginal levelized cost of energy bases optimal operation of distribution system considering photovoltaics, in *2017 North American Power Symposium (NAPS)* (2017), pp. 1–6
48. Available at: https://www.nrel.gov/analysis/tech-lcoe.html. Last Accessed: June 2020
49. Available at: https://www.nrel.gov/analysis/tech-cost-dg.html. Last Accessed: Mar 2021
50. W. Gutierrez, A. Ruiz-Columbie, M. Tutkun, L. Castillo, Impacts of the low-level jet's negative wind shear
51. J.S. Stein, C. Robinson, B. King, C. Deline, S. Rumme, B. Sekulic, *PV Lifetime Project: Measuring PV Module Performance Degradation: 2018 Indoor Flash Testing Results* (Sandia National Laboratories; National Renewable Energy Laboratory, Golden, CO, USA, 2018)
52. C. Roselund, Trackers dominate U.S. utility-scale solar (w/charts). PV Magazine, Sept 2017
53. C.D. Rodríguez-Gallegos, O. Gandhi, S.K. Panda, T. Reindl, On the PV tracker performance: tracking the sun versus tracking the best orientation. IEEE J. Photovolt. **10**(5), 1474–1480 (2020). https://doi.org/10.1109/JPHOTOV.2020.300699

Index

© The Author(s), under exclusive license to Springer Nature Switzerland AG 2022   51
S. Mil'shtein and D. Asthana, *Harvesting Solar Energy*, SpringerBriefs
in Materials, https://doi.org/10.1007/978-3-030-93380-7

Printed by Printforce, the Netherlands